2011 台达杯国际太阳能建筑设计竞赛获奖作品集
Awarded Works from International Solar Building Design Competition 2011

阳光与低碳生活
Low-carbon Life with Sunshine

中国可再生能源学会太阳能建筑专业委员会　编
Edited by Special Committee of Solar Buildings, CRES

执行主编　仲继寿　张磊
Chief Editor: Zhong Jishou, Zhang Lei

翻译　王岩
Translator: Wang Yan

中国建筑工业出版社
CHINA ARCHITECTURE & BUILDING PRESS

图书在版编目(CIP)数据

阳光与低碳生活/中国可再生能源学会太阳能建筑专业委员会编. —北京：中国建筑工业出版社，2011.6
2011台达杯国际太阳能建筑设计竞赛获奖作品集
ISBN 978-7-112-13240-9

Ⅰ.①阳… Ⅱ.①中… Ⅲ.太阳能住宅-建筑设计-作品集-中国-现代 Ⅳ.①TU29

中国版本图书馆CIP数据核字（2011）第088985号

本作品集由"台达环境与教育基金会"赞助出版

责任编辑：唐 旭 吴 绫
责任校对：关 健 王雪竹

2011台达杯国际太阳能建筑设计竞赛获奖作品集
Awarded Works from International Solar Building Design Competition 2011
阳光与低碳生活
Low-carbon Life with Sunshine
中国可再生能源学会太阳能建筑专业委员会 编
Edited by Special Committee of Solar Buildings, CRES
执行主编 仲继寿 张磊
Chief Editor: Zhong Jishou, Zhang Lei
翻译 王岩
Translator: Wang Yan

*

中国建筑工业出版社出版、发行（北京西郊百万庄）
各地新华书店、建筑书店经销
北京方舟正佳图文设计有限公司制版
北京中科印刷有限公司印刷

*

开本：787×1092毫米 1/12 印张：19 字数：543千字
2011年6月第一版 2011年6月第一次印刷
定价：88.00元
ISBN 978-7-112-13240-9
 （20681）

版权所有 翻印必究
如有印装质量问题，可寄本社退换
（邮政编码 100037）

地球是人类的共同家园，住宅是生活的基本场所，
太阳能是低碳住宅的永续动力，设计太阳能住宅，迈出低碳生活的第一步。
感谢台达环境与教育基金会资助举办2011国际太阳能建筑设计竞赛。
谨将本书献给低碳生活的设计者、建设者和践行者。

Earth is the common homestead of human beings. Dwelling house is the basic place we live. Solar energy is an eternal and continuous power to low-carbon dwelling house. Therefore, design of solar dwelling house is the first step for inspiriting low-carbon life.

Thanks to Delta Environmental & Educational Foundation that aids financially 2011 International Solar Building Design Competition.

This book is dedicated to the designer, builder and practitioner of low-carbon life.

目 录
CONTENTS

阳光与低碳生活　Low-carbon Life with Sunshine ... 008

过程回顾　General Background ... 009

2011台达杯国际太阳能建筑设计竞赛评审专家介绍
Introduction of Jury Members of International Solar Building Design Competition 2011 ... 016

一、综合奖作品 General Prize Awarded Works ... 001

一等奖　First Prize

垂直村落　Vertical Village ... 002

6米阳光　Share the Sunshine ... 006

二等奖　Second Prize

台院时光　Zero-carbon, Adaptable & Sustainable Residential Design ... 010

阳光·吴韵·绿宅　Sunshine Local Style Green House ... 014

园·自粉黛·光韵江南　The House Source from Jiangnan ... 018

光·波　Sunshine Wave ... 022

三等奖　Third Prize

享受绿色生活　Enjoy the Green Life ... 026

光·水谣　In-Light & Water ... 030

水·院·台　Water·Courtyard·Platform ... 034

光动能 Light Energy House	038
檐续曦望 Continuation of Hope	042
塞外风光 ECO Residence in Northern China	046

优秀奖 Honorable Mention Prize

阳光下的湖畔 Lakeside in the Sunshine	050
寻光之旅 In Search of the Way of Sunlight	054
太阳能花园 Solar Garden	058
富土阳光 Solar Energy Residence in Tongli	062
光合作用——阳光与低碳生活 Photosynthesis—Low-Carbon Life with Sunshine	066
阳光献礼 Solar Celebration	070
光之舞 Solar-House	074
阳光水榭 Sunny Waterside	078
土地"改革" Land Reform	082
地球24小时 Earth-24H	086
水意阳情 Solar & Sustainable Residence Design	090
山居 Shan Ju	094
阳光家园·江南情怀 Sunshine · Water · Home	098
沐姑苏 Livable and Low-carbon Residential Building Design in Wujiang	102
风光无限 Limitless Wind Limitless Solar	106

归园恬居 Low-carbon Residence Design	110
光·享住宅 Sunlight Enjoying Residence	114
二分宅 Brother House	118
阳光·风 Solar·Wind	122
灿阳千里 One Thousand Splendid Sun	126
新竹 The New "Bamboo"	130
把阳光带回家 Bring the Sunshine Home	134
向阳门第 Sun Mile	138
阳光、水、人家 Sunshine, Water, Home	142
阳光谷 Sunny Valley	146
忆江南 The Memory of Jiangnan	150
呼吸的蒙古包 Yurt Breathe	154
栖居·间奏 Dwells on This Earth·Rhythm	158
垂直阳光 Vertical Sunshine	162
吴江市低碳宜居住宅设计 Change in Sunshine	166

二、技术专项奖作品　Prize for Technical Excellence Works　171

Ice-ray Apartment	172
阳光住宅 Solar House	176

| 光·的容器 Solar · Vessel | 180 |
| 三五间舍 Three or Five Houses | 184 |

参赛人员名单　Name List of Attending Competitors
188

2011台达杯国际太阳能建筑设计竞赛办法
Competition Brief for International Solar Building Design Competition 2011
196

后记　Postscript
208

阳光与低碳生活
Low-carbon Life with Sunshine

  世界气候变暖，一次性能源的消耗与日俱增，人类的生存面临着空前的挑战。越来越多的人认识到，只有采用低碳的生活方式，充分利用太阳能等清洁能源，建设低能耗的宜居住宅，摒弃高舒适度等于高能耗的错误住宅建造理念，才能将节能减排的目标落到实处，从而改善人居环境，实现低碳减排目标。本届竞赛的主题为"阳光与低碳生活"，面向全球组织作品征集，汇集全球更多设计团队的智慧，为将太阳能建筑从蓝图转为现实作出贡献。竞赛吸引了国内外共1550个团队注册，征集有效作品188份。国际太阳能建筑设计竞赛是配合世界太阳能大会两年一届定期举行的常规赛事，与一般竞赛仅停留在理念上不同，把"阳光照进现实"是此赛事一大亮点。2009国际太阳能建筑设计竞赛的一等奖作品"蜀光"已经在四川省绵阳市杨家镇台达阳光小学实施，并投入使用。同样，本届竞赛的获奖作品也将在美丽的江南水乡——苏州同里湖畔进行实地建设。

  竞赛不是目的，以竞赛的形式推动太阳能等清洁能源在建筑、生活中的应用，传播环保节能理念，以使我们在寻找替代性的清洁能源和技术方案的同时，改变人们的能源消费习惯和观念，建立一种节能、环保、可持续的生活方式，这才是竞赛的内涵所在。我们欣喜地看到，国际太阳能建筑设计竞赛已经成为具有一定品牌影响力的建筑设计活动，一批年轻的建筑师脱颖而出，通过竞赛平台走上工作岗位，在推广低碳绿色节能建筑方面崭露头角。更多落地实践的优秀作品将通过对未来建筑的无限思考，建筑今天的精致生活。感谢2011台达杯国际太阳能建筑设计竞赛活动的参与者，感谢所有关心与支持太阳能建筑发展的人们。

  科技引领时代，创意点亮未来。从我做起，从我们的住宅做起，把太阳能利用纳入到建筑中来，与建筑技术融为一体，让温暖的技术塑造人们未来的生活空间。低碳生活，从居住开始。

  The climate of the world is going warmer and warmer and one-off consumption of the energy is increasing day by day. Now human being's survival is facing an unprecedented challenge. More and more people have realized that none but adopting the life style with low carbon, fully utilizing solar energy and other clean energy resources, constructing livable dwelling house with low energy consumption and getting rid of wrong conception of housing construction that high comfort is equal to high energy consumption, the aim of saving energy and decreasing emission will be really implemented, thus improving residential environment of people and realizing low carbon and emission decrease.

  The subject of current building design competition is "Low-carbon Life with Sunshine". It collected works from all over the world, focused on the wisdom of more design teams to make the contribution of solar buildings from design blueprint into realty. The competition attracted 1550 teams from domestic and abroad to register and 188 effective works were collected. The competition is a conventional one taken place every two years parallel with the Solar World Congress. One of its light points different from other competitions that only rest on conception is letting sunlight shining into the reality. First prize awarded work "First Light of Morning" in 2009 International Solar Building Design Competition has been constructed and put in use in Yangjia Zhen, Mianyang City, Sichuan Province named Delta Sunlight School. The awarded work of the current competition will also be built near the beautiful lake of Tongli.

  The competition is not the purpose. Its meaning in intension is just to promote the application of solar energy and other energy sources to buildings and life, promulgate the conception of environmental protection and saving energy in order to change the custom and idea about energy consumption and set up an energy efficient, environmental friendly and sustainable way of the life while search replaceable clean energy sources and technical schemes. We have delightedly seen that International Solar Building Design Competition has been a building design activity with a certain brand influence. A passel of young architects have stood out, gone to work via the competition platform and make their figure on popularizing low carbon, green and energy efficient buildings. More architectural works will be put into implementation and set up exquisite life at present through deep consideration for intending architecture.

  Thanks to all participants of 2011 Delta Cup — International Solar Building Design Competition. Thanks to the all who concern and support the development of solar buildings.

  Science and technology are leading the time. Originality will emblaze the future. Let's start from ourselves and from our dwelling houses to put solar application into buildings and integrate it into building technology. Let the warm technology create life space for people's future.

  Low-carbon life starts from inhabitation!

过程回顾
General Background

竞赛场地实景　Site of the Competition

　　本次竞赛由国际太阳能学会和中国可再生能源学会联合主办；国家住宅与居住环境工程技术研究中心、中国可再生能源学会太阳能建筑专业委员会承办；台达环境与教育基金会独家冠名。在各单位的通力协作下，竞赛组委会于2010年1月成立，并组织了竞赛启动、媒体宣传、校园巡讲、作品注册与提交、作品初评与终评、技术交流会等一系列活动。这些活动得到了海内外业界人士的积极响应和参与，尽管太阳能住宅的设计难度大，依然有188位参赛作者为我们呈上了精彩的答卷。

一、竞赛筹备

　　集合住宅具有用户产权分散、使用工况多样、涉及技术全面的特点，一直是实施太阳能综合利用较为困难的建筑类型。住宅是我们生活的场所，也是实践低碳生活方式的重要载体，为了推动太阳能等清洁能源在住宅中的应用，竞赛组委会决定迎难而上，将本届的竞赛题目确定为太阳能住宅。

　　自筹备之初，竞赛组委会积极地与各方联系，寻求获奖作品实地建设的场地。在台达环境与教育基金会、吴江经济技术开发区管理委员会的大力支持下，

The competition is organized by International Solar Energy Society as well as Chinese Renewable Energy Society (CRES). The Competition operators are China National Engineering Research Center for Human Settlements and Special Committee of Solar Buildings, CRES and the sponsor is Delta Environmental & Educational Foundation. Based on full cooperation of all relevant organizations, the Organization Committee of the competition was set up in January 2010 and it organized a series of activities such as competition startup, media publicity, circuit lecture in campus, registration and submission of works, preliminary and final evaluation of the work, technical seminar, etc.. These activities gained active response and participation of professionals domestic and abroad. Although solar dwelling house is difficult to design, there are still 188 participants submitted their excellent works.

1 Competition Preparation

Collective house is such a building type difficult for integrated application of solar energy because the property rights of the owners are scattered, the situation of building use is various and all aspects of technology are involved.

组委会最终选定在江苏省吴江市，美丽的同里湖畔进行实地建设。通过组织专家进行实地考察，确定了设计竞赛的场地建设条件，并编制了竞赛设计任务书。

二、竞赛启动

2010年6月23日，2011台达杯国际太阳能建筑设计竞赛在北京启动。针对住有所居、节能减排的需求，本次竞赛以"阳光与低碳生活"为主题，面向全球组织作品征集，竞赛题目分为吴江市低碳宜居住宅和呼和浩特市低碳宜居住宅两项，参赛人员可任选一项进行设计。打造以太阳能利用为主的低碳宜居住宅，在教育、环保与大众领域全力推广低碳生活理念，让太阳能建筑的设计理念深植在未来建筑师的心中。

发布会出席嘉宾留影　Guests in the competition conference

三、校园巡讲

自竞赛启动以来，组委会向国内37所高校发送了以往竞赛的获奖作品展板和本届竞赛的宣传资料；经过周密的筹划与准备，竞赛组委会组织了为期一个月的"2011台达杯国际太阳能建筑设计竞赛"校园巡讲活动。2010年12月1日在同济大学首站开讲，其后又分别走进了东南大学、大连理工大学、清华大学、北京工业大学、哈尔滨工业大学、华中科技大学等院校传播太阳能建筑设计理念，2010年

Dwelling house is the place where we live and important carriers to implement low carbon life style. In order to popularize the application of solar energy and other clean energy resources on buildings the Organization Committee of the competition decided to go forward against difficulties and made solar dwelling house as the subject of the competition.

Since the beginning of the preparation, the Organization Committee started to actively get in touch with all aspects to search a site for the implementation of awarded work. With the support of Delta Environmental & Educational Foundation and Management Committee of Wujiang Technology Development Area, finally it was decided that the project will be constructed by the lake in beautiful Tongli, Wujiang City, Jiangsu Province. Through a review on the spot by experts, the site environment and design conditions are made sure and design assignment was worked out subsequently.

2 Competition Start-up

2011 Delta Cup - International Solar Building Design Competition was started up in Beijing on 23 June, 2010. The theme of the competition is "Low-carbon Life with Sunshine" and the works were collected from the whole world. Two subjects of design for choice: low- carbon livable dwelling house in Wujiang and in Huhhot. Participants could choose one of them to design. Constructing low carbon livable dwelling house mainly depending on using solar energy and fully popularizing the conception of low-carbon life in the field of education, environmental protection and to inhabitants, the design idea of solar building will be deeply implanted into hearts of future architects.

3 Circuit Lecture in Campus

Since competition start-up, the Organization Committee has sent show boards about awarded works in previous sessions and information of current competition to 37 domestic universities and colleges. By means of thorough plan and preparation, the Organization Committee has also taken a circuit lecture concerning 2011 competition during a month. The lecture about design idea of solar buildings was taken place in Tongji University firstly on 1 December, 2010 and then in Southeast University, Dalian University of Technology, Tsinghua University, Beijing University of Technology, Harbin Institute of Technology and

巡讲地图　Map of circuit lecture

东南大学巡讲现场　Circuit lecture in Southeast University

华中科技大学巡讲现场
Circuit lecture in Huazhong University of Science & Technology

西北工业大学巡讲现场　Circuit lecture in Northwest Polytechnical University

大连理工大学巡讲现场　Circuit lecture in Dalian University of Technology

12月29日校园巡讲在西北工业大学落下帷幕。巡讲的内容包括太阳能建筑应用现状、发展前景及往届竞赛获奖作品介绍等，均受到了在校师生的热烈欢迎。注册人数在巡讲后明显上升。

四、媒体宣传

在专业技术巡讲的同时，自2010年6月至2011年1月，组委会开展了多渠道的媒体宣传工作，包括：竞赛双语网站实时报道竞赛进展情况并开展太阳能建筑的科普宣传；在百度刊登关键字搜索广告，以便社会大众更快捷地登陆竞赛网站；在中国《建筑学报》、《建筑技艺》、美国《AR》、韩国《SPACE magazine》、

Huazhong University of Science and Technology. The circuit lecture has rung down the curtain in Northwest Polytechnical University on 29 December, 2010. The lecture including existing situation of solar buildings, developing prospect and introduction of awarded works in previous sessions was warmly welcome by teachers and students. After the lectures, the registrations were creasing evidently.

4 Publicity through the Media

At the same time of circuit professional lecture, the Organization Committee launched a media campaign via diverse channels from June 2010

百度搜索引擎推广 Search engine on BAIDU website

在国内外网站上的报道 Website report domestic and abroad

ISES 的官方网站 Official website of ISES

日本《A+U》等近10家国内外专业杂志刊登了竞赛活动宣传专版；在雅虎网、新华网、ABBS、筑能网、日本AKICHIATLAS.com、美国ArchNewsNow.com、意大利architecture.it、丹麦bygnet.dk等国内外70余家网站上报道或链接了竞赛活动相关信息；在《科技日报》、《中国建设报》、《环球时报》等30余家国内的平面媒体，向国际发布了竞赛的组织与宣传情况。

五、竞赛注册及提交情况汇总

本次竞赛的注册时间为2010年6月25日至2011年1月20日，共1550个团队注册，其中境外注册团队11个，涵盖日本、印度、法国、印度尼西亚、韩国等国家和我国港澳台地区。截至2011年3月1日，竞赛组委会共收到国内外所提交的作品191项，其中有效作品188项，无效作品3项。有效作品中涉及吴江地区的作品136项，涉及呼和浩特地区的作品52项；作者依然以国内建筑院校为主，共提

to January 2011 including real-time report about the process of the competition and popularization on solar buildings through a two-language website, an advertisement of key word search published on BAIDU website so that people could land competition website quickly, a special space of the report for competition activities on about ten professional magazines domestic and abroad such as "Architecture Journal" and "Construction Craft" of China, "AR" of USA, SPACE magazine of Korea, "A+U", etc., report and relevant information about the activities linked on more than 70 websites such as Yahoo, Xinhua, ABBS, topenergy.org, AKICHIATLAS.com (Japan), ArchNewsNow.com (USA)、architecture. it (Italy), bygnet.dk (Demark), etc. and report about the organization and publicity of the competition to the world published on more than 30 print medias such as Science & Technology Daily, China Construction, Globe Times and others.

交183项。

六、作品初评

2011年3月5日，组委会将全部有效作品提交给初评专家组。每位专家根据竞赛办法中规定的评比标准对每一件作品进行评审，各自选出60份进入终评的作品。经过竞赛评审专家的严格审查，3月20日组委会对所有专家的评审结果进行统计后，获得评审专家半数以上投票的、共52项作品进入终评阶段。

七、作品终评

2011年台达杯国际太阳能建筑设计竞赛终评会于2011年4月6～7日在江苏省吴江市同里召开。经专家组讨论，一致推选国际太阳能学会亚太区主席、澳大利

5 Competition Registration and Works Submission

Registration period of the competition is from 25 June, 2010 to 20 January, 2011. In this duration 1550 teams have registered. 11 teams of them are abroad from Japan, India, France, Indonesia, Korea and other countries as well as those from the areas of Hong Kong, Macao and Taiwan. Up to 1 March, 2011 the Organization Committee has received 191 works from home and abroad, 188 works are effective and 3 works are invalid. In effective works 136 are for Wujiang area while 52 are for Huhhot area. Most authors are from domestic architecture universities and colleges which submitted 183 works.

6 Preliminary Evaluation

On 5 March, 2011 the Organization Committee submitted all of effective works to experts for appraisal. Every expert gave evaluation to every work

终评会现场　Scenes of final evaluation conference

讨论作品　Discussion

终评专家组合影　Members of final evaluation jury team

亚新南威尔士大学建筑环境系Deo Prasad教授担任本次终评工作的评审组长。在他的主持下，评审专家组按照简单多数的原则集体讨论和公正客观地评选作品，通过三轮的投票与评论，最终选出2项一等奖作品（吴江及呼和浩特地区各一件）、4项二等奖作品、6项三等奖作品以及30项优秀奖作品；技术专项奖与建筑创意奖共4项。

八、组织国际可再生能源建筑集成技术交流会

2011年4月6日上午，由中国可再生能源学会太阳能建筑专业委员会与吴江经济技术开发区管理委员会共同在吴江市同里湖大酒店举办了2011国际太阳能建筑技术交流会。交流会邀请国内外的太阳能建筑专家与当地的设计单位、太阳能生

according to the appraisal standard of competition regulation and chose 60 works for final evaluation. On 29 March, the Organization Committee conducted a statistical review to initial evaluation and 52 works were voted out for final evaluation, which got more than half of the votes of all experts.

7 Final Evaluation

Final evaluation of 2011 Delta Cup – International Solar Building Design Competition was taken place in Tongli, Wujiang on 6-7 April, 2011. Through a discussion in expert group, Mr. Deo Prasad, Asia-Pacific President of International Solar Energy Society, Professor of the Faculty of the Building

交流会现场　Scenes of the seminar

产企业、相关研究机构、建设主管单位就国内外太阳能建筑的应用现状、发展趋势及优秀的工程案例进行交流。以竞赛获奖作品在吴江的实地建设为契机，借鉴发达国家的成功经验，推动太阳能等清洁能源在吴江，在江南，甚至在国内的推广应用。

Environment, University of New South Wales, Sydney, Australia was elected as the director of the jury team for final appraisal. Under his presidency, in accordance with the principle of simple majority, jury team members discussed together, objectively & fairly evaluated and selected with three round voting and appraisal. Finally, awarded works were selected out. They are: Two works of First Prize (One of them is in Wujiang area and another in Huhhot area), four works of Second Prize, Six works of Third Prize and 30 works of Honorable Mention Prize. Besides, there are still four works of Prize for Technical Excellence and Architectural Originality.

8 International Communication Seminar on Building Integrated Renewable Energy Technology

In the morning on 6 April, 2011, International Solar Building Technical Communication Seminar was taken place at Tongli Lake Hotel hold by Special Committee of Solar Buildings, CRES and Management Committee of Wujiang Economic and Technical Development Area. Experts from domestic and abroad as well as local design institutes, solar industries, relevant research institutes and construction administration were invited to do intercourse concerning existing situation of solar application in buildings, developing prospect and excellent projects. With the construction project based on first prize work in Wujiang as an opportunity and using successful experiment of advanced countries, solar energy and other new energy sources will be popularized and adapted progressively in Wujiang, southern China and even the whole country.

基地实地考察

2011台达杯国际太阳能建筑设计竞赛评审专家介绍
Introduction of Jury Members of International Solar Building Design Competition 2011

评审专家:
Jury Members:

Peter Luscuere: 荷兰代尔伏特大学（TU Delft）建筑系教授。
Mr. Peter Luscuere: Professor of Department of Architecture, TU Delft, The Netherlands.

Deo Prasad: 国际太阳能学会亚太区主席、澳大利亚新南威尔士大学建筑环境系教授。
Mr. Deo Prasad, Asia-Pacific President of International Solar Energy Society (ISES) and Professor of Faculty of the Built Environment, University of New South Wales, Sydney, Australia.

Mitsuhiro Udagawa: 国际太阳能学会日本区主席、日本工学院大学建筑系教授。
Mr. Mitsuhiro Udagawa, President of ISES-Japan; Doctor from Engineering of Waseda University and Professor of Department of Architecture, Kogakuin University.

M.Norbert Fisch: 德国不伦瑞克理工大学教授（TU Braunschweig）、建筑与太阳能技术学院院长。
Mr. M.Norbert Fisch, Professor of TU Braunschweig, President of the Institute of Architecture and Solar Energy Technology, Germany and Doctor from Stuttgart University, Germany.

崔恺： 国际建筑师协会副理事、中国建筑学会副理事长、中国国家工程设计大师、中国建筑设计研究院副院长、总建筑师。
Mr. Cui Kai, Deputy Board Member of IUA (International Union of Architects); Vice President of Architectural Society of China; National Design Master and Chief Architect of China Architecture Design & Research Group.

仲继寿：中国可再生能源学会太阳能建筑专业委员会主任委员、国家住宅工程中心主任。
Mr. Zhong Jishou, Chief Commissioner of Special Committee of Solar Building, Chinese Renewable Energy Society and Director of China National Engineering Research Center for Human Settlements.

冯雅：中国建筑西南设计研究院副总工程师、中国建筑学会建筑热工与节能专业委员会副主任。
Mr. Feng Ya: Deputy Chief Engineer of China Southwest Architectural Design Institute, Deputy Director of Special Committee of Building Thermal Engineering and Energy Saving, China Architectural Society.

喜文华：甘肃自然能源研究所所长、联合国工业发展组织国际太阳能技术促进转让中心主任、联合国可再生能源国际专家、国际协调员。
Mr. Xi Wenhua: Director of the Institute of Natural Energy Resources of Gansu Province, Director of Promotion and Transfer Center of International Solar Technology, United Nations Industrial Development, International Specialist of Renewable Energy, UN and International Coordinator.

黄秋平：华东建筑设计研究院副总建筑师。
Mr. Huang Qiuping, Vice-chief Architect of East China Architecture Design & Research Institute.

林宪德：台湾绿色建筑委员会主席、台湾成功大学建筑系教授。
Mr. Lin Xiande, President of Taiwan Green Building Committee and Professor of Faculty of Architecture of Success University, Taiwan.

叶青：深圳市建筑科学研究院有限公司党委书记、董事长。
Ms.Ye Qing, Secretary of Party and Chairwoman of the Board Committee of Shenzhen Institute of Building Research.

一、综合奖作品
General Prize Awarded Works

一等奖
First Prize

项目名称：垂直村落
Vertical Village

作　者：顾雨拯、季鹏程、
　　　　　杨维菊

参赛单位：东南大学建筑学院

专家点评：

作者从中国书法大师的绘画中寻求灵感，将太阳能利用与江南民居的设计理念有机结合，整体设计理念新颖，住宅平面设计基本合理。作品在建筑的南立面上充分实现了主被动太阳能的利用，包括建筑遮阳、被动通风与太阳能集热，为实现太阳能综合利用与建筑一体化方面开拓了新的思路。不足之处在于设计对太阳能技术的细节阐述得不够详细。

Gaining inspiration from paintings of Chinese calligraphist, the designers have made solar application a nice combination with local housed of southern China. The idea of integrated design is novel and plan design of the housing building is rational basically. On the south facade of the building active and passive applications of solar energy are realized including sunlight shelter, passive ventilation and solar energy collection. It carves out a new thoughtfulness to realize comprehensive application of solar energy and building integration. It is inadequate that a particular description about solar technology is not fully given out.

太阳能住宅设计——垂直村落

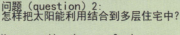
VERTICAL VILLAGE

问题（question）1:
怎样将水乡肌理反映到现代多层建筑中？
How to reflect the texture of local watery region in modern multi-storey？

问题（question）2:
怎样把太阳能利用结合到多层住宅中？
How to combine the use of solar energy with the multi-storey disign？

回答（answer）:
在吴冠中的画中，我们找到了解答。
We find the answer in the traditional Chinese paintings created by Wu Guanzhong.

国画中用物体的上下排布表达远远的透视关系，我们将之运用于建筑设计中：一个垂直的村落。波浪形倾斜的墙的运用使每一户都同时获得了向阳面和遮阳面，提高了太阳能热水器的得光率，同时形成的自遮阳解决了南向眩光问题。

We choose the concept of *vertical village* in our design by realizing the relationship between different objects in traditional Chinese paintings which show the depth through arranging them up and down. The use of undulating tilted wall make it possible for each house get both a sunny side and a shading side. In other words, this operation not only raises the utilization rate of solar water heater but also avoids the problem of south glare.

太阳能住宅设计 —— 垂直村落
VERTICAL VILLAGE

2011 台达杯国际太阳能建筑设计竞赛获奖作品集

BRIDGE

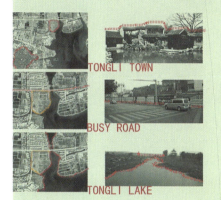
TONGLI TOWN
BUSY ROAD
TONGLI LAKE

Economic and technical norms:
Site area: 11000m²
Construction area: 17495m²
Floor area ratio: 1.59
Greening rate: 40.62%
Number of households: 144
Ground parking: 15

SITE PLAN 1:500

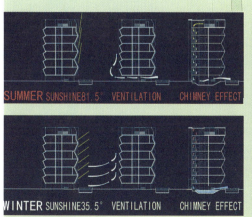
SUMMER SUNSHINE 81.5° VENTILATION CHIMNEY EFFECT
WINTER SUNSHINE 35.5° VENTILATION CHIMNEY EFFECT

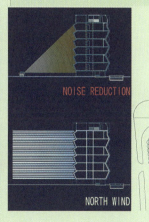
NOISE REDUCTION
NORTH WIND

UNDERGROUND PARKING
NOISE REDUCTION

SOUTH ELEVATION 1:200

WEST ELEVATION 1:200

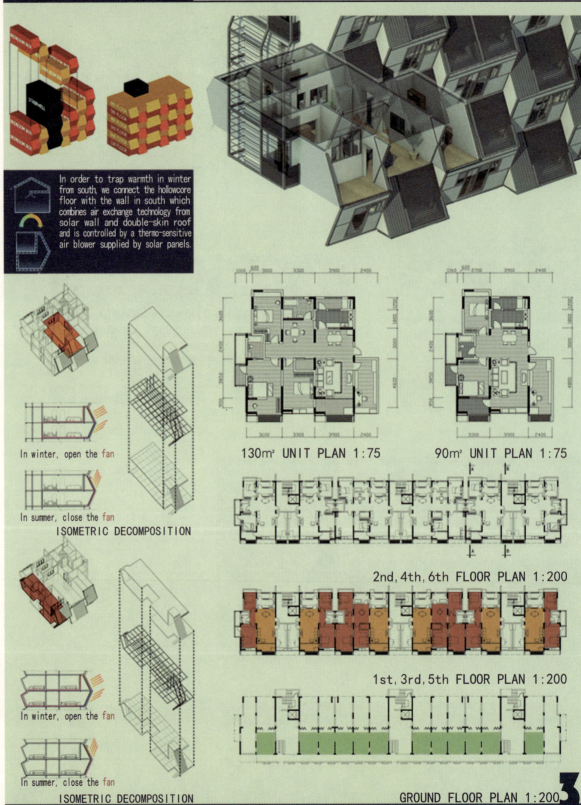

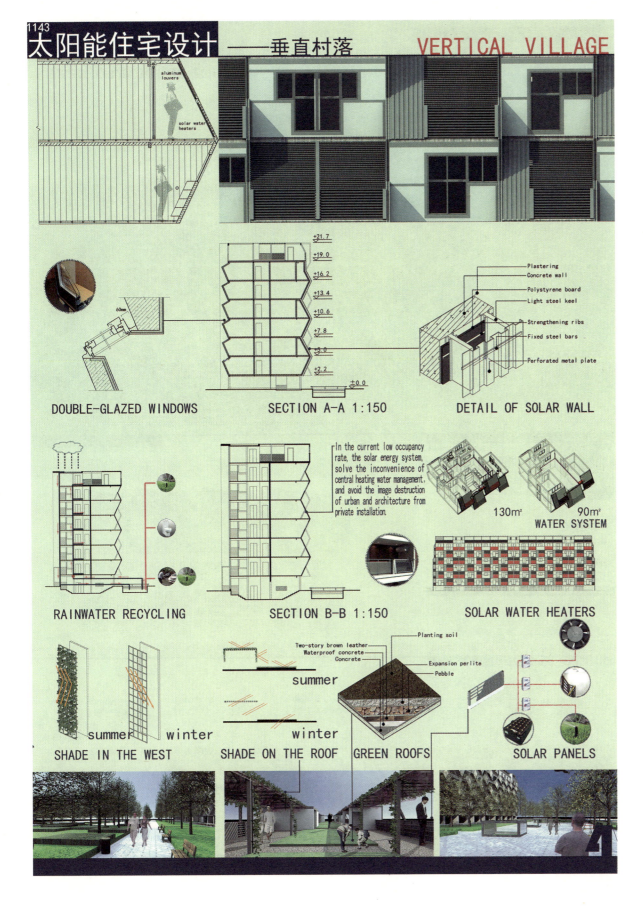

一等奖
First Prize

项目名称：6米阳光
Share the Sunshine

作　者：夏海山、姜忆南、杜晓辉、
　　　　王　佳、刘冬贺、孟璠磊、
　　　　侯　磊

参赛单位：北京交通大学建筑与艺术系

专家点评：

该设计的平面非常规整，通过与阳光室结合的公共空间营造邻里交往空间，较好地解决了北方小面积户型设计问题；通过模拟软件进行项目预分析，得出科学合理的建筑朝向、遮阳、声环境等；在此基础上合理利用主、被动太阳能技术及其他节能技术，提高了南北房间风、光、热的能量交换效率。

The plan designed is very regular. It solves the problem concerning design of small dwelling unit in northern China very well by means of making public space combined with solar room as a neighbor intercourse room. A pre-analysis of the project by stimulant software shows rational orientation, sunlight shelter and sound environment of the building, based on which active and passive solar energy technology and other ones of energy saving are applied, thus upgrading the efficiency of energy exchange of wind, light and heat.

SHARE THE SUNSHINE
6米阳光

HOHHOT RESIDENTIAL DESIGN

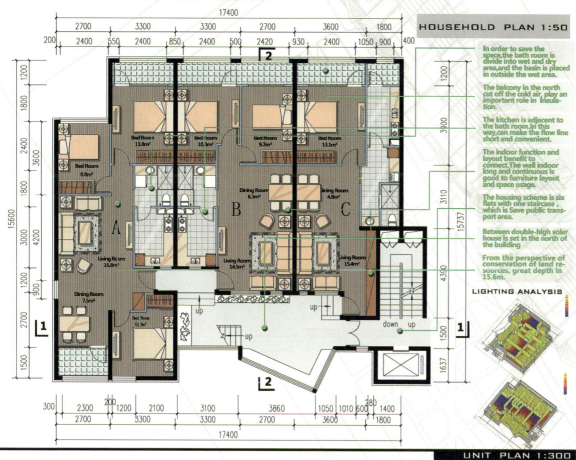

HOUSEHOLD PLAN 1:50

- In order to save the space, the bath room is divide into wet and dry area, and the basin is placed in outside the wet area.
- The balcony in the north cut off the cold air, play an important role in Insulation.
- The kitchen is adjacent to the bath room. In this way, can make the flow line short and convenient.
- The indoor function and layout benefit to connect. The wall indoor long and continuous is good to furniture layout and space usage.
- The housing scheme is six flats with one staircase, which is Save public transport area.
- Between double-high solar house is set in the north of the building.
- From the perspective of conservation of land resources, great depth In 15.6m.

LIGHTING ANALYSIS

UNIT PLAN 1:300

Apartment A
Usable area : 74.4 m²
Construction area : 95.0 m²
Using the coefficient : 0.82

Apartment B
Usable area : 52.3 m²
Construction area : 65.0 m²
Using the coefficient : 0.81

Apartment C
Usable area : 50.1 m²
Construction area : 63.6 m²
Using the coefficient : 0.80

HOUSEHOLD MODEL

	A	B	C
Type	1B	2B	3B
Usable	74.4 m²	52.3 m²	50.1 m²
Construction	95.0 m²	65.3 m²	63.6 m²
Construction (per floor)	498.2		
Pool area (per floor)	29.8		
Using coefficient	0.81		
Total floor area	21912.8		

FLOOR PLAN 1:300

FIRST FLOOR

TYPICAL FIRST FLOOR

TYPICAL SECOND FLOOR

TOP FLOOR

SOLAR ENERGY ARCHITECTURE DESIGN

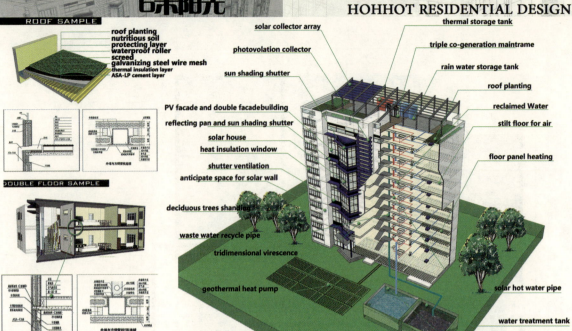

SHARE THE SUNSHINE 6米阳光

HOHHOT RESIDENTIAL DESIGN

CROSS SECTION

section 1-1 1:300   section 2-2 1:300

Thermal Analysis

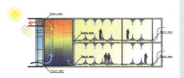

winter Day
Sun-light heat the air in solar-house in the south of building. Under the action of the, the warm air is transported to the north of buiding through the double-floor, completing the thermal cycling.

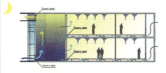

Winter night
Solar-house and heat storaging wall absorb the heat during the daytime, and release them at night. Basically keep the status quo to maintain the indoor temperature.

Summer day
During the daytime, the wind gap is open to guide the natural ventilation pass through. At the same time, the solar-house is playing an important role in sun-shading system. The solar-house collects heat and help make air flow in the room.

summer night
At summer night, the wind gap of solar-house is open. The warm air is transported to the north of building through the double-floor, completing the thermal cycling.

Sunshade Analysis in Summer

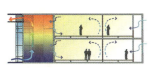

In summer day, the solar altitude is high. The adjustable shutter sun-shading system out of solar-house, avoids in the sun-light perpendicular incidence room.

Lighting Analysis in Winter

Put on cover and open shutter to make light go straight to bottom of the atrium. The angle of the refraction of the sunlight is desiged according to the local latitude. It can reflect sunlight in summer.

Ventilation Analysis

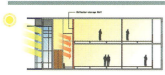

Summer
Wind gap of solar-house is open to rapid flow of air in the south, driving the air movement.

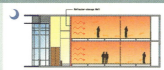

Winter
Wind gap of solar-house is closed warm the cold air and make indoor air exchange.

Heat storaging wall

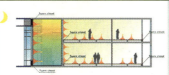

Daytime
The wall using heat storaging material can collect heat in daytime.

Nightime
Heat storaging wall radiate heat at night, making the air indoor warm and comfortable.

PERSPECTIVE

SOLAR ENERGY ARCHITECTURE DESIGN

二等奖
Second Prize

项目名称：台院时光
Zero-carbon, Adaptable & Sustainable Residential Design

作　　者：王　瑞、谷亚兰、莫颖媚、王国栋

参赛单位：西南交通大学建筑学院

专家点评：

设计通过建筑底部二层与上部四层的错位设计，形成底层架空公共空间。太阳能板与立面结合，形成灰白相间、错落有致、富有江南韵味的建筑意象。主被动太阳能利用的技术设计表达完整，在太阳能建筑集成利用方面具有可操作性。

In the scheme, an open public space is formed by dislocation design between lower two stories and upper four stories. The appearance has the style of local house in southern China by alternate arrangement of gray and white on facade, which is a good combination of solar boards. Technical design about active and passive applications of solar energy is described perfectly and it has operability.

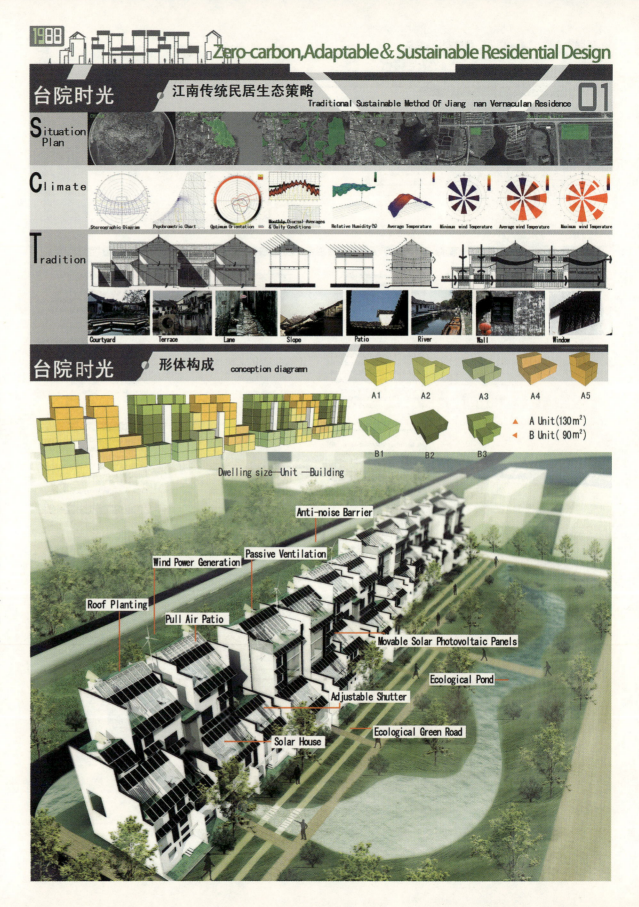

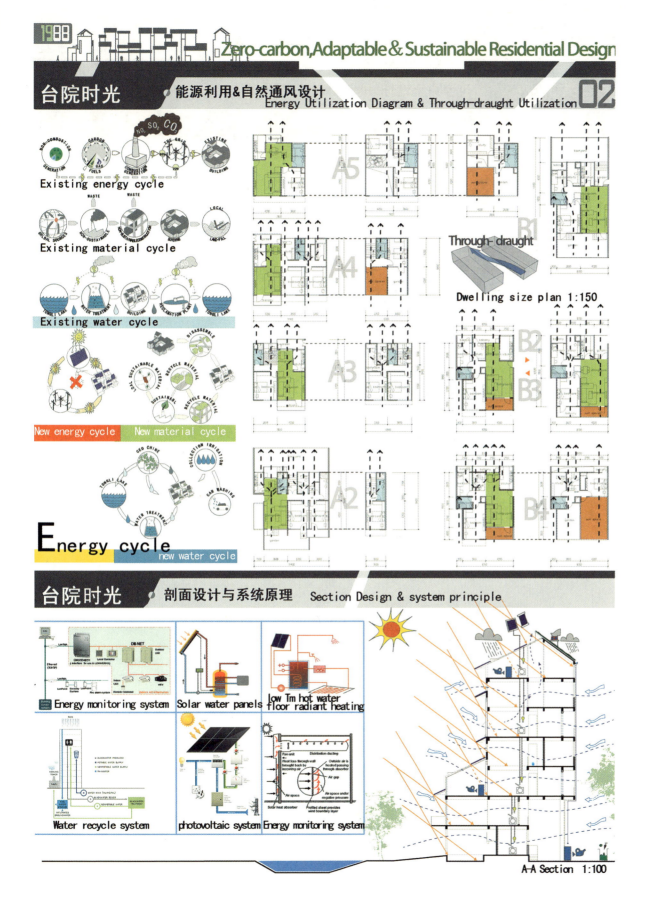

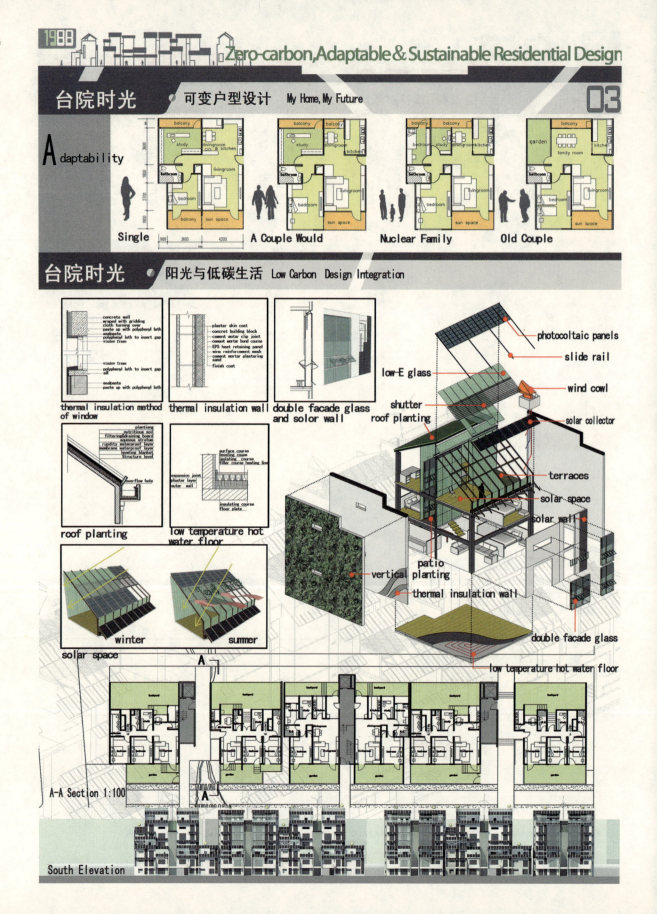

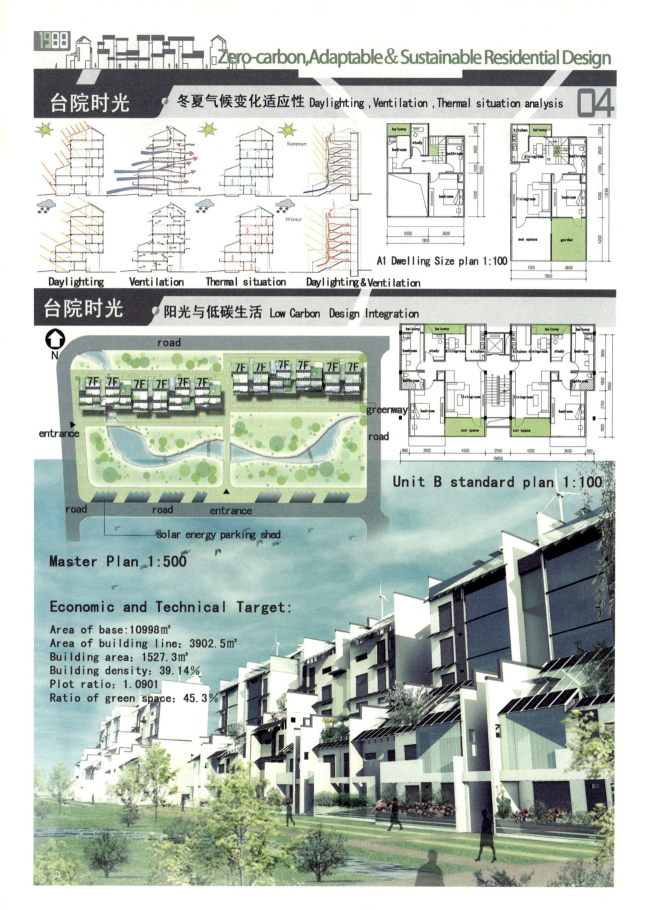

2011台达杯国际太阳能建筑设计竞赛获奖作品集

二等奖
Second Prize

项目名称：阳光·吴韵·绿宅
Sunshine Local
Style Green House

作　者：江文婷、蔡权、刘磊、孙湉

参赛单位：南京工业大学建筑学院

专家点评：

建筑造型富有江南地方民居特色，光伏屋顶整体感强，庭院设计与建筑群体关系良好。作品较好地将被动与主动太阳能利用技术与建筑集成，通过室内热环境的分析体现出绿色节能的理念。

The form of the building is designed with the character of local houses in southern China and PV system roof has a good integrated sense. The scheme is good at the application of active and passive solar energy technology and building integration as well. It embodies green concept of energy saving via the analysis on inner heat environment.

DESIGN SPECIFICATION

The project is based on ecological design. First of all, the site was reasonably layouted. We increase the distance between the two rows of buildings furthest in order to increase daylight hours for the buildings in the north. The design ensures comfortable living conditions and provides the possibility of evolution for different families.

On the part of building energy conservation, according to local climate and the difference of the energy consumption in winter and summer, the design consists in passive solar energy utilization, trys to utilize other solar energy technology combine with the architectural style. At the same time, we pay attention to the practicability and the economy of the technology.

设计说明：

本方案以"阳光低碳生活"为主题，以住宅的生态化设计为出发点，首先对场地进行了合理的规划布局，最大限度地增加南北两排建筑之间的距离以增加北面建筑的日照。住宅建筑设计首先保证户型平面的舒适性，同时为不同住户或同一住户的不同时段提供户型演变的可能性。

在建筑节能上，根据当地气候特点以及夏季和冬季住宅能耗的不同，设计以被动式太阳能技术为主，并结合造型尝试多项太阳能技术的综合运用，同时注重技术的适用性和经济性。

NO. 2014

HOUSE TYPE ECONOMIC INDEX						
HOUSE	TYPE	NUMBER	PROPORTION (%)	NET AREA (m²)	GROSS AREA (m²)	NET AREA FACTOR (%)
A	A1	30	41.7	66.8	90.32	73.96
	A2	6	8.3	66.8+23.85=90.64 including loft	90.32+30.6=120.92 including loft	74.90
B	B1	30	41.7	97.65	132.13	73.90
	B2	6	8.3	97.65+35.57=133.22 including loft	132.13+46.4=178.53 including loft	74.62

1F PLAN 1:200

阳光·吴韵·绿宅 Sunshine Local Style Green House 02
"Low-Carbon Life with Sunshine" 2011 Delta-cup International Solar Building Design Competition

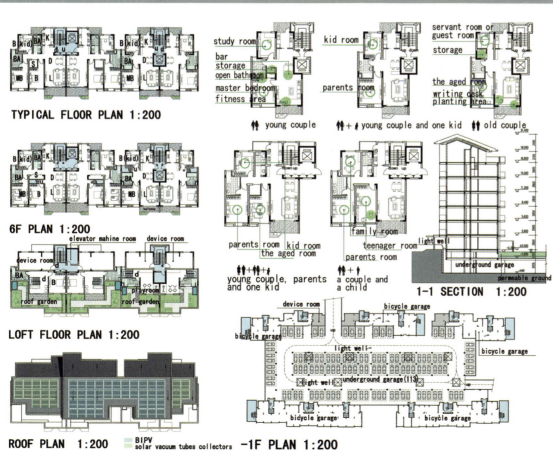

TYPICAL FLOOR PLAN 1:200

6F PLAN 1:200

LOFT FLOOR PLAN 1:200

ROOF PLAN 1:200
■ BIPV
■ solar vacuum tubes collectors

-1F PLAN 1:200

1-1 SECTION 1:200

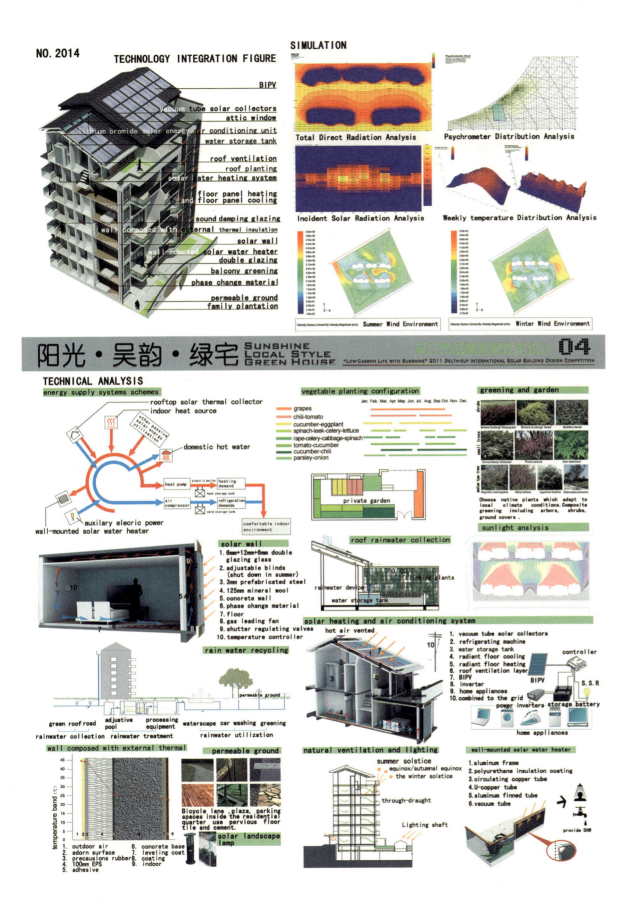

二等奖
Second Prize

项目名称：园·自粉黛·光韵江南
The House Sourse From Jiangnan

作　者：高力强、姚　瑶、姚辰明、
　　　　　王　帅、刘　恋、沈纪超、
　　　　　欧阳仕淮、宋宏浩

参赛单位：石家庄铁道大学建筑与艺术学院

专家点评：
作品利用现代居住建筑特点，创新性地提出建筑层级空间概念，建筑体量适宜，层次感强，具有较强的江南建筑特色，深具都市美学印象。设计对太阳能、风能及植被进行合理的应用，通风楼板应用概念新颖。

The work embodies the character of modern housing building and puts forward the concept of building space with various levels. The building is with a proper mass and the layout has a layered arrangement. The design also has the character of local houses in southern China and produces a vivid impression of city esthetics. Solar energy, wind energy and vegetation have been rationally utilized. The idea of ventilation floor slab is novel.

区位分析 / Site Analysis

    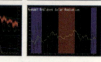 

Basic issue: Thousand-year-old town style continuing
Basic solution: The continuation of texture. The utilization of traditional strategy

Basic issue: High temperature in summer while low in winter. High humidity
Basic solution: Thermal insulation. Dehumidification

效果图 / Perspective

生态思考 / Ecological Thinking

Natural Resources | Technical Measures

Vegetation Effect | Architecture Form

Ecological Strategy
Season's changing, how can the apartment respond with suitable utilization of solar technology?
With the replacement of night and day, how to ensure the comfort of the apartment?

Sloping Roof | Courtyard
Horsehead Walls | Traditional Strategy
How can the traditional strategy combined with modern ones to adapt to local climate?

园·自粉黛·光韵江南
THE HOUSE SOURCE FROM JIANGNAN
LIFE ECOLOGICAL TRADITIONAL

01

The Eco-House Design in Wujiang

总平面图 / General Layout 1:500

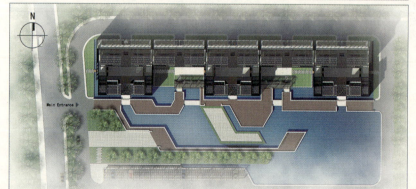

生态策略 / Ecological Strategy

Vertical Solar Energy Schematic
- Photovoltaic System
- Solar Collector
- Sun-shading System
- Refractive Device

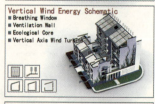
Vertical Wind Energy Schematic
- Breathing Window
- Ventilation Wall
- Ecological Core
- Vertical Axis Wind Turbine

Vertical Virescence Space Schematic
- Garden Afforestation
- Balcony Afforestation
- Ecological Core Afforestation
- Western Wall Afforestation
- Roof Afforestation

设计说明
本案位于千年古镇同里。通过对传统民居村落的解读，方案延续了古镇"小桥，流水，人家"的规划特征和"民风多古朴，住宅尽清幽"的建筑特点，同时结合现代住区元素，引入层级空间概念，形成公共院落、入户平台、私家阳台、屋顶花园等多强度等级空间构成的庭院式建筑格局。人们可以交往也可以独处，归属感和私密感同时存在。
方案学习借鉴了传统民居的生态管理思想，并注重对太阳能、风能和植被的控制和利用。将传统生态策略和新技术整合于居住建筑可持续性设计之中，实现对阳光低碳生活的地域性回答。

Design Description
The project, located in Tongli, is thousand-year-old. The dwelling continues the characteristics of living by water and peaceful architectural features of primitive simplicity. Meanwhile, combined with modern residential elements, the housing introduce hierarchy space concept, shaping publicyard, platform, balcony, roof garden, grades of space composing traditional court-yard building structure. People can have intercourse and be alone, with the protection of ownership and privacy.
Plan learn from the traditional local-style dwelling houses of the ecological philosophy thoughts, with emphasize on solar, wind, vegetation control and utilization. The traditional ecological strategy and new ones integrate in residential building sustainable design to realize low-carbon life with sunshine of regional answer.

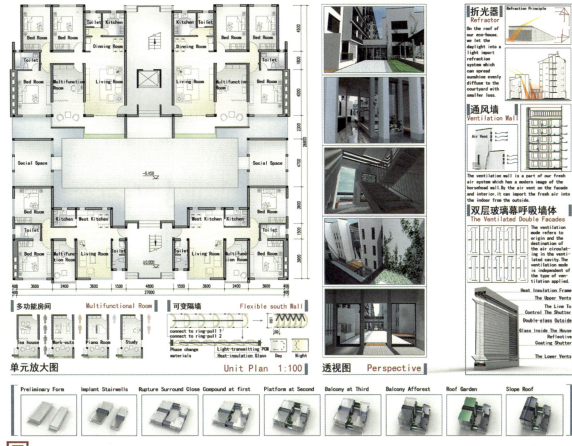

THE HOUSE SOURCE FROM JIANGNAN
LIFE ECOLOGICAL TRADITIONAL

02

生态屋面分析　Eco-roof Analysis

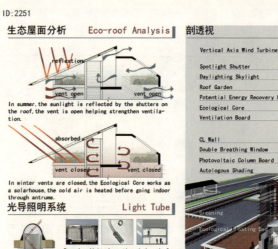

In summer, the sunlight is reflected by the shutters on the roof, the vent is open helping strengthen ventilation.

In winter vents are closed, the Ecological Core works as a solarhouse, the cold air is heated before going indoor through antrums.

剖透视　Section Perspective

- Vertical Axis Wind Turbine
- Spotlight Shutter
- Daylighting Skylight
- Roof Garden
- Potential Energy Recovery Elevator
- Ecological Core
- Ventilation Board
- CL Wall
- Double Breathing Window
- Photovoltaic Column Board
- Autologous Shading
- Greening
- Ecological Heating Bed
- Rain Water Recycling

光导照明系统　Light Tube

By using light-absorbing shade and efficient light transmission system. It can provide good illumination for the space where is lacks of delighting, therefore making the interior lighting environment more natural and comfortable.

CL墙体　CL Wall

Lightweight partition
- Gypsum stud
- 12~20mm fibrous plaster
- Acoustic material
- 12~20mm fibrous plaster

Main wall
- Plasterboard
- Concrete Layer
- Insulation layer
- Concrete layer

100x100 steel wire mesh

Advantages: CL walls applies to multilayer residence, with strong thermal and sound insulation performance. CL walls realize construction component factory and material recycling, which helps save energy and cultivated land.

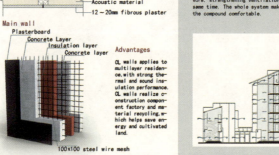

Ventilation Analysis In Summer
Having been cooled down by the pond, Summer-led winds go through the ecological core, bringing fresh air in. Wind deflectors in the north form negative pressure, strengthening ventilation at the same time. The whole system make life in the compound comfortable.

Ventilation Analysis In Winter
To make shelter from the wind, wind deflectors are closed during winter, thus the Ecological Core works as a solar house. The Ecological Core warm the air before it going indoor. Meanwhile the closed core protect the environment in the compound against northwest wind.

Daylighting And Energy Analysis
In summer, the sun's elevation angle's large. The reflector helps reflect the sun's light onto out with ventilation cool down indoor temperature. In winter by the calculation of solar elevation angle, the sunlight directly splaying stack. The air is heated by sun and ventilated by thermal pressure.

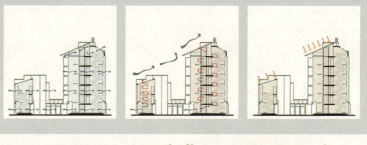

Solar & Wind Energy Source
Surface-water Source Heat Pump & Fresh Air System
Artificial Wetland & Sewage Treatment

园·自粉黛·光韵江南

THE HOUSE SOURCE FROM JIANGNAN
LIFE · ECOLOGICAL · TRADITIONAL

03

The Eco-House Design in Wujiang
Building With Carbon Green

院落生活　Life In Compound

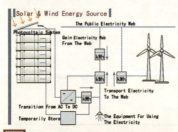

用绿色的思维建新建筑

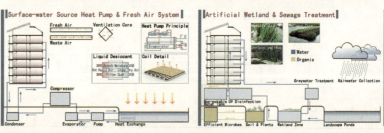

Multistoreyed Apartment: Plate housings are unable to creat space of the franchisee, which lead to less communication between neighbors.

Compound Apartment: Organizing surround close to the compound provides the opportunity for neighborhood talk.

Ecological Compound: Combine ecological strategy with the compound to improve living environment through low carbon ways.

Multistoreyed Apartment: Linear plan make residential area lack of vigor and outdoor space play traffic role only.

Compound Apartment: Compound residential building produce nodes, forming settlements to public space hierarchy.

Ecological Compound: Introduce Ecological Core to improve residential micro climate, making the compound more suitable for life.

二等奖
Second Prize

项目名称：光·波
　　　　　Sunshine Wave
作　　者：高力强、刘　恋、沈纪超、
　　　　　王　帅、欧阳仕淮、
　　　　　宋宏浩、张袆娇、徐禛龙
参赛单位：石家庄铁道大学建筑与艺术学院

专家点评：

M形住宅楼栋平面设计适合西北气候特点，夏季南北通风，冬季南向围合空间少受西北风侵袭，北向有一半的房间避开了西北或东北风。建筑与庭院结合良好，住宅平面功能良好，凸显北方住宅保温防风特色。

Housing building plan of M type is suitable to the climate of northwest in China, so it has north-south ventilation in summer and enclosed space toward south by buildings may avoid the attack of northwest wind in winter. Half of rooms toward north will keep away from northwest or northeast wind. The buildings have a good combination with court yard. The function of plan is better and it displays the character of housing building in northern China in keeping warm and preventing cold wind.

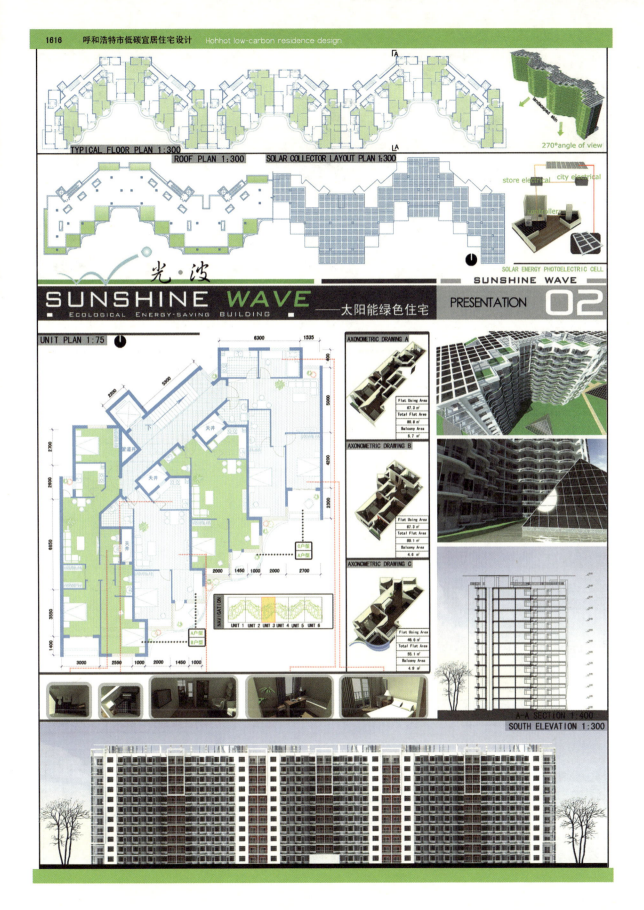

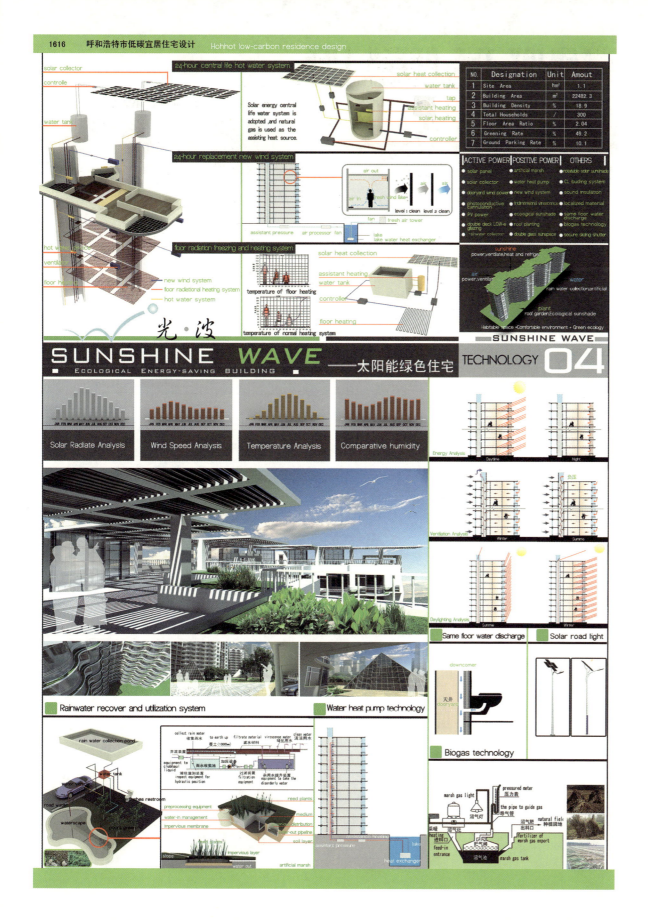

三等奖
Third Prize

项目名称：享受绿色生活
Enjoy the Green Life
作　者：张　伟、吴　悠、杨维菊
参赛单位：东南大学建筑学院

专家点评：
方案整合了住宅阳台的构造元素与太阳能集热器，从而形成了丰富有趣的立面肌理效果。内天井的设计具有创新，但也带来了视线干扰等其他问题。

There is a good integration of housing balconies and solar energy collectors, thus forming an interesting texture on facade. Inner courtyard is an innovated design; however, it also brings some problem of disturbance on line of sight and others.

1109 ENJOY THE GREEN LIFE
THE LOW-CARBON AND LIVABLE DESIGN FOR WUJIANG DWELLING

MATERIAL USED

The site of this project is located in TONGLI. The architectural style is as the picture represented

Grey Tiles

Blue Bricks

White Walls

FACADE DESIGN

Traditional Residence Facade

No Sun-shading System.
WE SAY NO TO THIS
WE NEED A NEW ONE

New facade we designed

Providing Self-shading system.

SPACE DESIGN AND GREEN STRATEGY

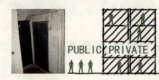
Traditional buildings cut off the links between the residents.

CUT AND BREAK

Our design provide the residents a semi-public atrium.

INTRODUCE SUNLIGHT AND WIND

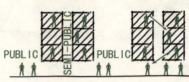
The atrium can also let the sunlight and the wind in.

01

设计说明

在空间上，本设计旨在打破中国旧有住户与户之间相互隔离的状态，给住户提供一个中庭空间以利于相互交流，恢复旧有的邻里模式；在生态上，利用中庭产生热力拔风的效应，增强自然通风，并结合地源热泵，提供一个舒适的室内环境。同时，在立面设计上使用自遮阳的概念，体现绿色、生态的设计理念。

In space design, we break the old domestic buildings and households state of isolation between users, giving tenants an atrium space to facilitate mutual exchange;Ecologically, the use of heat generated in the atrium of Stack effect increasing strong natural ventilation, provide a comfortable indoor environment. Meanwhile, in the design of the facade we use the self-shading system, reflects the green and ecologydesign concept.

ANALYSIS OF GENERAL PLAN 1:800

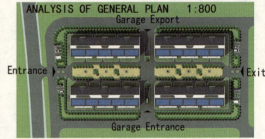

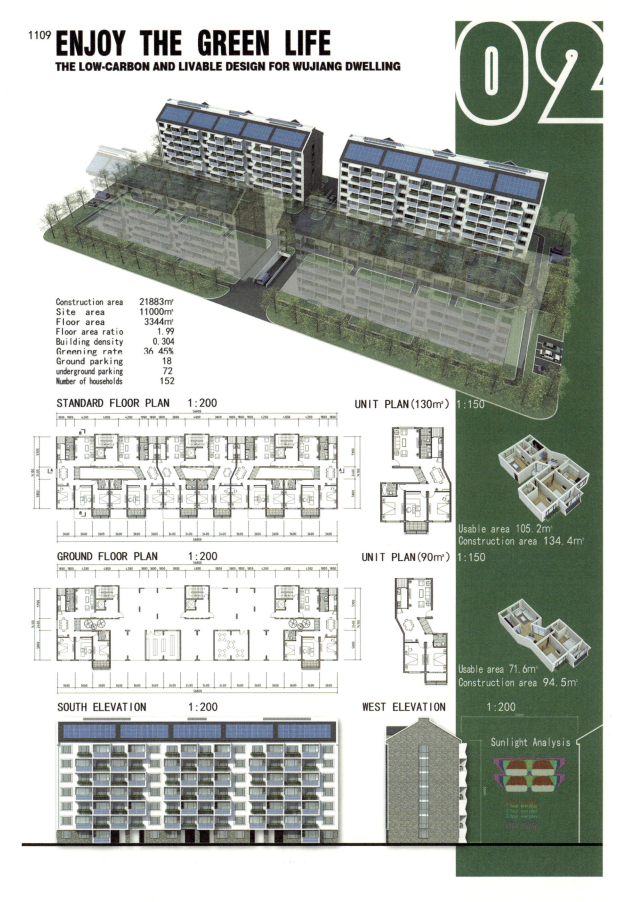

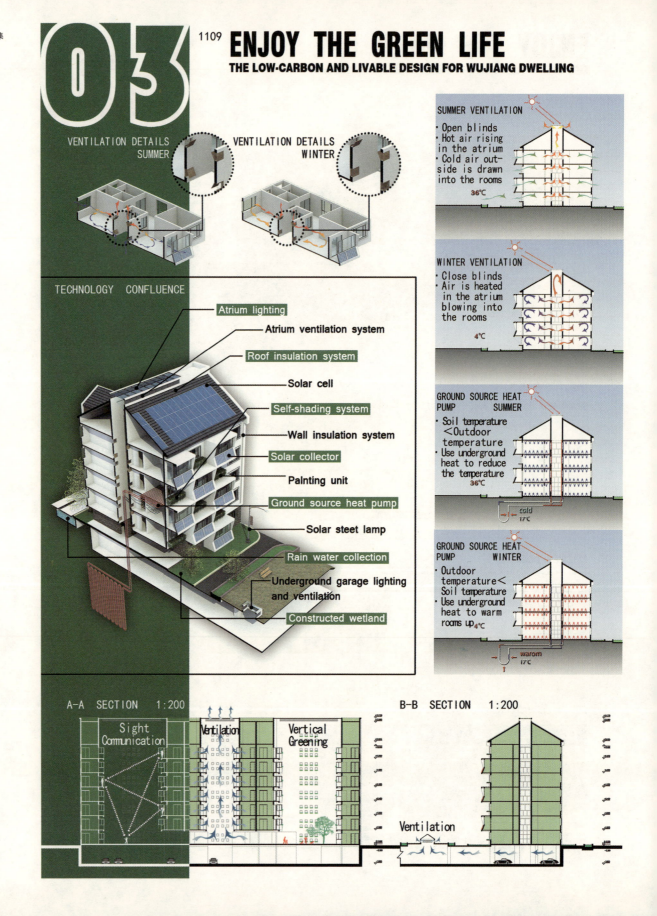

1109 ENJOY THE GREEN LIFE
THE LOW-CARBON AND LIVABLE DESIGN FOR WUJIANG DWELLING

RAIN WATER COLLECTION

- Solar cell
- Rain water
- Electricity
- Irrigation
- Washing cars
- Irrigation
- Flushing
- Available water
- Waste water
- Water pump
- City water system

HOUSE WATER CYCLE SYSTEM

- residential water pipe network
- water meter
- solar water heater
- residential rainwater recycling

SHADING SYSTEM
- No Shading Measures
- Self-Shading
- Balcony-Shading
- Roller Shutter Sun Shading system

SOLAR WATER CYCLE SYSTEM

Cold / Hot

SECTION DETAIL 1:25

- roofing tile
- 1:3 cement sand rendering (with reinforcing rod)
- 1:2 cement sand rendering
- waterproof layer
- 1:2 cement-sandmix Toweling Course
- basic unit roof
- insulation board
- 1:2 cement sand rendering
- sink drain
- eavesgutter
- gargoyle
- aluminium panel
- laminate flooring
- (GSHP) flexible pipe
- 1:2 cement sand rendering
- floorslab
- 1:2 cement sand rendering
- solar water heater
- balcony
- transparent elastic putty
- rendering coat mortar
- alkaline-resistant fiberglass roving cloth
- external coating
- cement sand rendering
- rigid insulation board
- adhesive
- basic unit wall
- cement sand rendering
- window superior volumes boxes
- sun-shading rolling shutter
- XPS
- interior
- heat insulating window
- indoor floor
- underground garage
- ground floor
- vapour barrier
- ceiling of the basement
- waterproof concrete slabs
- 40mmC20 fine gravel concret
- Asphalt paper isolation layer
- 1.5mmLB-16PVC waterproof roll
- Cement mortar screed-coat
- C20 precast concrete plank
- C10 concrete cushion
- foundation

04

2011 台达杯国际太阳能建筑设计竞赛获奖作品集

三等奖
Third Prize

项目名称：光·水谣
In-Light & Water
作　　者：李晓东、岳文昆、朱　堃、
　　　　　武鼎鑫、夏晨晨、黄　莹
参赛单位：东南大学建筑学院

专家点评：
作品建筑设计合理，能较好地利用建筑水环境，改变建筑微气候。太阳能板（或集热器）与建筑的集成匹配了当地建筑的大众色彩，浅层地热能的利用具有可操作性，同时覆土导风坡、渗水路面等也是作品的亮点。不足之处在于太阳能通风塔的通风性能有限，还有可能会造成噪声干扰。

The design is reasonable and good at utilizing water environment for the building, thus improving micro-climate. Solar board (or solar heat collector) is integrated with building and suitable to popularized colors of local buildings. The application of terrestrial heat on flat layer is operable. Also there are some light points such as leading wind sloping field and penetrable pavement of road. It is inadequate that the performance of solar ventilation tower is limited with possibility of noise interruption.

光·水谣 ARCHITECTURE & LANDSCAPE
In-Light & Water 江苏吴江同里古镇低碳宜居住宅方案设计

[A] Base Analysis

[Location Analysis of Historical & Cultural Background]

Tongli Town is located in Wujiang, Jiangsu Province, 80km away from Shanghai and 20km from Suzhou. It has a long history and a typical style of water town. With its beautiful scenery, the town is surrounded by water. There are 7 isles within the town, separated by 15 rivers and connected with each other by 49 bridges. It has a reputation as the "Oriental Venice" in the world for its bridges over flowing streams, located beside the Taihu Lake. It is surrounded by Shanghai, Suzhou, Hangzhou, which are three famous cities in southern China. The total area of the town is 133.15 square kilometers and the population is 55,000.

[Environmental Analysis & Climate Characteristics]

Tongli, located in the north subtropical monsoon climate region, is warm, humid and rainy, obvious in monsoon and four seasons—winter and summer long, spring and autumn short. The average frost-free period can up to 233 days. There are a variety of unique microclimates due to the differences in topography, latitude and other factors within the territory.

The Base is located in Millennium town, first by the cultural atmosphere and architectural style of southern external effects, followed as close to the same Lake, southeast to the King to the good, East and southeast winds to the surface of Residential Group will be an important factor in comfort. The west side of the road shared with the Lake Hotel, and the north side of the road for the city main road, northwest of noise would be more serious.

[Climate maps]

[B] Actuality of Bvases

The base is located in the northwest corner of the Tongli Lake, 200 meters away from the Lake. The base has an already built residential area in the west and is close to the historical and cultural town-Tongli to its south. Its transport is convenient.

[C] Technology Showcase

A Solar chimny
B Solar panels
C Solar wall
D Windowsill plant tank
E Adjustable shade
F Wells through cold
G Variable ecological pond
H Solar collector
I Wind channel
I The front overhead bottom
J Underground garage
L Heat Exchanger
K Artificial wetland degradation

【PERSPECTIVE】

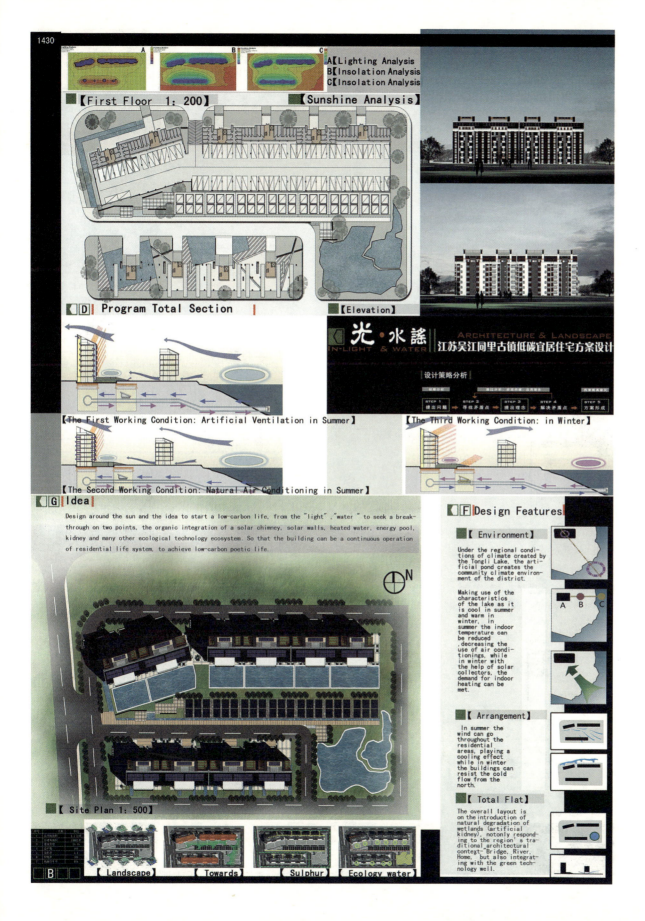

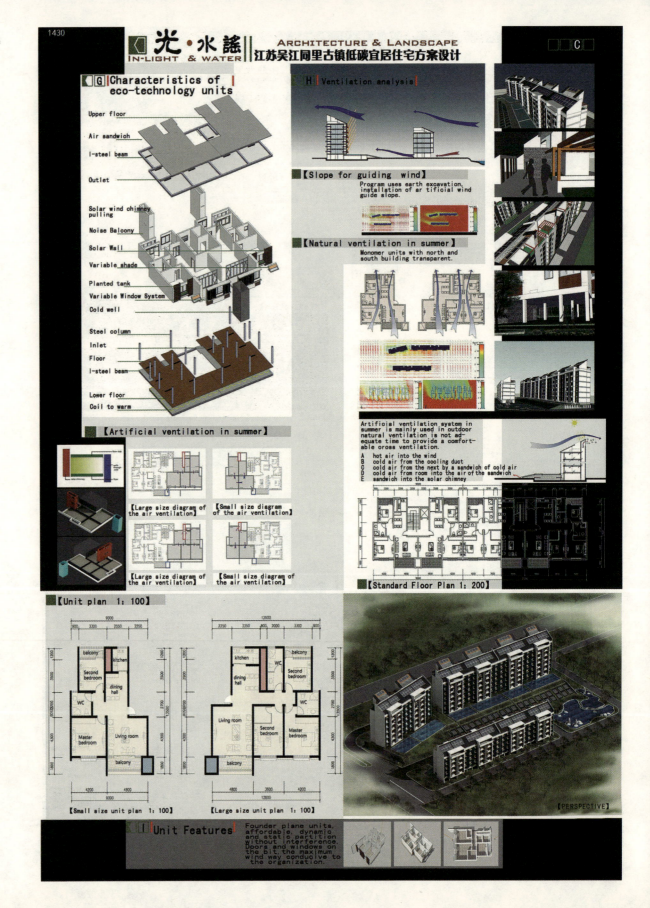

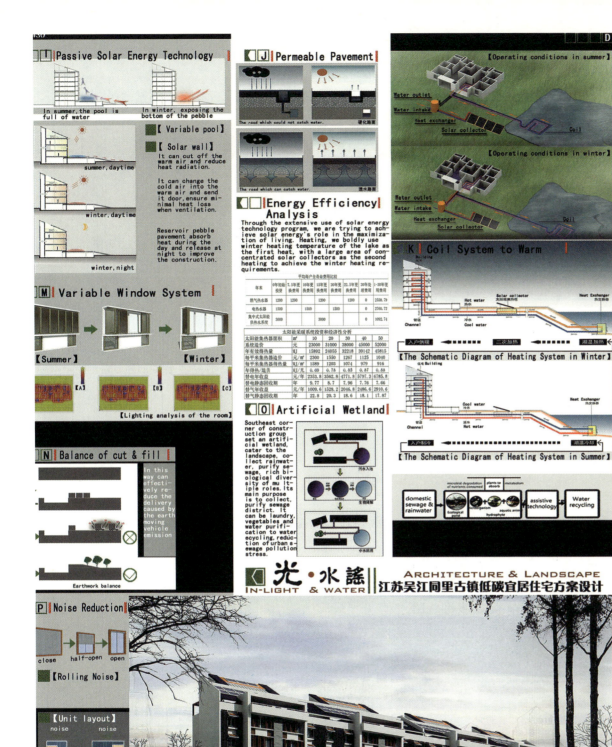

三等奖
Third Prize

项目名称：水·院·台
　　　　　Water·Courtyard·Platform
作　　者：魏瑞涵、张晨光、王寅璞、
　　　　　刘滨洋、姜咏茜、杨星晰
参赛单位：山东建筑大学建筑城规学院

专家点评：

作品具有水乡建筑的风格与形态，平面布局合理，空间组合灵活。多种模拟软件的分析为太阳能建筑利用提供了较好的设计依据。结合庭院设计的底层南入口空间具有层次感，从底层到六层的退台式平面处理，形成了丰富的建筑立面。太阳能技术应用较为合理。

The work has style and form of water village. Plan arrangement is reasonable and space composition is flexible. The analysis of various stimulant software gives more data for solar application. The space of south entry on first floor designed with courtyard together has layered sense. Balconies stand back step by step from first floor to sixth floor and make the facade abound in change. It is reasonable in application of solar technology.

LIVABLE RESIDENTIAL BUILDING DESIGN — LOW-CARBON DELTA-CUP 2011
LOW-CARBON AND LIVABLE RESIDENTIAL BUILDING DESIGN

[INTRODUCTION]

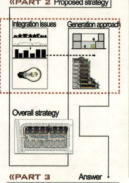

《PART 1 — Design mission / Research
《PART 2 — Proposed strategy — Integration issues / Generation approach / Overall strategy
《PART 3 — Answer

水·院·台
吴江市同里古镇低碳宜居住宅建筑设计　[阳光住居]·[低碳生活]
[Low-Carbon Life with Sunshine]
[DELTA-CUP INTERNATIONAL SOLAR BUILDING DESIGN]

[DESIGN NOTES]

This project's site is located in Wujiang city, on the north-west of Tongli Lake bank.
We introduce the water from Tongli lake to the site to conduct our site planning. The concept of integration of architecture design and the technology design is conducted. We inherit the culture of courtyard, terrace and sloping roof. The courtyard provide the residents place of living and leisure, and the terrace create a good indoor and outdoor transitional space.

基地位于吴江市内，毗邻风景秀丽的同里湖，具有典型的江南水乡风貌。
基地的规划设计从水乡建筑布局及形态入手，延续水系脉络并与场地设计和建筑节能相结合。设计融合了传统民居的庭院、阳台等空间形式。坡屋顶，庭院可以作为室外起居空间，退台形成良好的室内外过渡空间。

JIANGSU / WUJIANG

[CONTRADICTIONS]

Traditional residential buildings built along the lake. | Urban grow big, broad roads extend in the water region. | The relationship between residence and environment become tough. | The city expand rapidly, bulky buildings seize the conventional district and the environment.

[URBAN]

The original scale and texture of the city disappearance of it, is replaced by the almost bulky boxes, the cities in the world looks the same gradually.

[COMMUNICATION]

 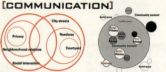

The disappearance of the original courtyard and street, resulting in people communicate rarely.

[BACKGROUND]

Water town in the south of Yangtze River is beautiful, but style of recent years is the destruction of rivers and lakes.

[STRATEGY]

Depressive space scale is not conducive for people to sharing ideas. | Flexible design of space to ease the repression. | Interesting design with more terrace space. | Multivariate space is conducive for people to exchange and enjoy the landscape.

1.

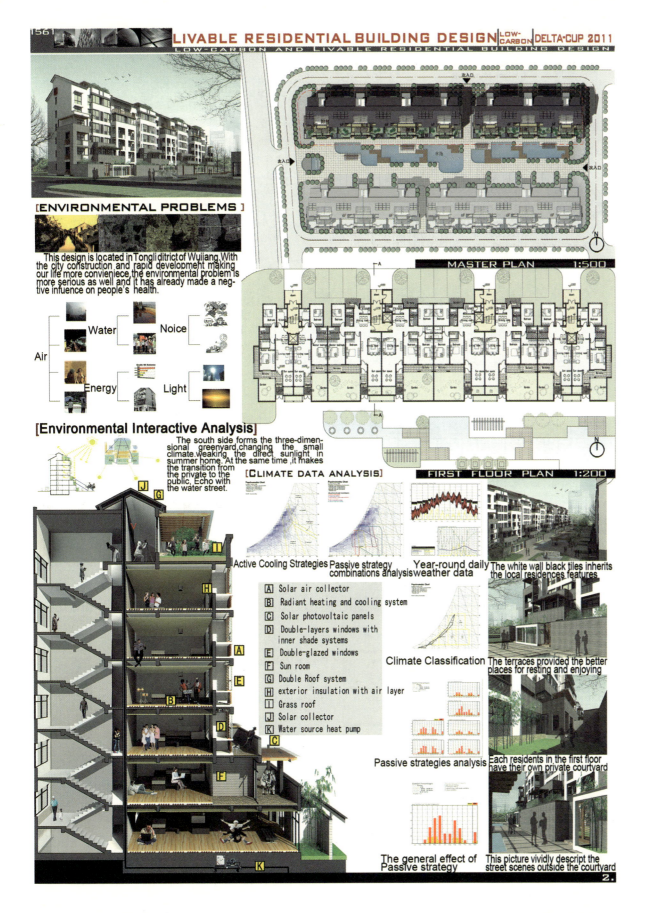

LIVABLE RESIDENTIAL BUILDING DESIGN | LOW-CARBON | DELTA-CUP 2011
LOW-CARBON AND LIVABLE RESIDENTIAL BUILDING DESIGN

[UNITS ANALYSIS]
This type of units has open space, bright living room, square kitchen with natural light and ventilation. The active spaces are divided distinctly from the quiet space.

[130 AQUARE UNITS]
This type of units has compact size, transparent living room, bright bedrooms, and square kitchen. There is spacious terrace in the south where you can watch the water landscape view.

[90 AQUARE UNITS]

[SOLAR AIR COLLECTOR]
- outer glass
- laminated louver
- inner glass
- grid solar air heater

[EXTERIOR INSULATION WITH AIR LAYER]
- primary wall
- insulation
- metal keel
- clay plate

The wall that prevent the cold wind from the outer side.

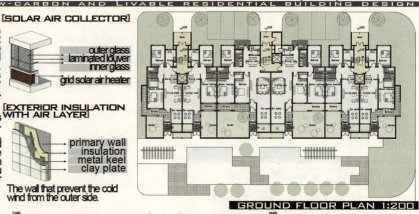

GROUND FLOOR PLAN 1:200

SECOND FLOOR PLAN 1:300

SIXTH FLOOR PLAN 1:300

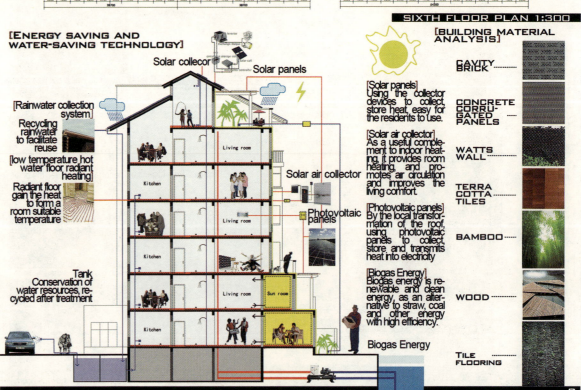

[ENERGY SAVING AND WATER-SAVING TECHNOLOGY]

[Rainwater collection system]
Recycling rainwater to facilitate reuse

[low temperature hot water floor radiant heating]
Radiant floor gain the heat to form a room suitable temperature

Tank
Conservation of water resources, recycled after treatment

Solar collector
Solar panels
Solar air collector
Photovoltaic panels
Biogas Energy

[BUILDING MATERIAL ANALYSIS]

[Solar panels]
Using the collector devices to collect, store heat, easy for the residents to use.

[Solar air collector]
As a useful complement to indoor heating, it provides room heating, and promotes air circulation and improves the living comfort.

[Photovoltaic panels]
By the local transformation of the roof, using photovoltaic panels to collect, store and transmits heat into electricity

[Biogas Energy]
Biogas energy is renewable and clean energy, as an alternative to straw, coal and other energy with high efficiency.

- CAVITY BRICK
- CONCRETE CORRUGATED PANELS
- WATTS WALL
- TERRA COTTA TILES
- BAMBOO
- WOOD
- TILE FLOORING

LIVABLE RESIDENTIAL BUILDING DESIGN — LOW-CARBON DELTA-CUP 2011
LOW-CARBON AND LIVABLE RESIDENTIAL BUILDING DESIGN

[LIGHTING ANALYSIS]

SUMMER: Solar elevation angle is relatively large, eaves and cantilevered balcony in the south can effectively reduce the radiation and the grille of the high window reflects sunlight and prevent the light from shining through.

WINTER: By the calculation of solar elevation angle, the Cornice and balcony of the sun does not prevent the light from shining through, and because of the paralleled direction of the grid with the angle of the sunlight, the light can direct into the room.

[Economic and technological indexes]

Number	Items	Unit	Quantity
1	Total land area	Hectares	1.1
2	Total construction area	m²	8865.46
3	Building density	%	16.41%
4	Total households	Household	84
5	Cubage rate	—	0.81
6	Greening rate	%	32%
7	Ground parking rate	%	19%

[Shadings, Overshadowing and Sunlight Hours]

Fan-shaped envelope body shadow analysis shows a window is kept out of sunshine by other buildings, thus adjustment buildings planning to meet sunlight requirements.

According to the sunshine duration analysis results we can adjust building span that meets the two hours of sunlight on the winter solstice.

[SPLIT COMPONENT ANALYSIS]

[Thermal analysis]

WINTER: • sunlight direction • water pipe radiation

SUMMER: • radiant cooling

[Ventilation Analysis]

ACTIVE: • treat the temperature and humidity • guarateen fresh air and exhaust used air

PASSIVE: • natural ventilation • chimney effect

SOUTH ELEVATION | WEST ELEVATION | A-A SECTION

三等奖
Third Prize

项目名称：光动能
　　　　　Light Energy House
作　　者：葛贵武、林波荣、吴锡嘉、
　　　　　李　娴、彭　渤
参赛单位：解放军后勤工程学院军事建筑
　　　　　工程系、清华大学建筑学院、
　　　　　广西艺术学院设计学院

专家点评：
作品结合当地传统建筑风格，较好地应用了被动设计策略。智能幕墙系统和可变发射率的围护结构设计新颖，建筑群体与景观配置良好。

The work is better in adopting strategy of passive design combined with traditional style of local houses. Intellective curtain wall system and inclosing structure design with changeable emissivity is novel. The layout of building group and sight are better.

2272

组群 One

Light energy house 光动能

[区位介绍]
Location Description

Wujiang is located in the far south of Jiangsu Province, China, right at the golden triangle intersection of Jiangsu, Zhejiang and Shanghai. The site is located in Wujiang, north of "historical and cultural city"tongli, and near West tongli Lake Bank. Its north of the city is main roads—Jiangxing Dong Road.

Location map

[技术指标]
onotechnical norms

No	Name	Unit	Remarks
1	total land area	3890m²	
2	total building area	8724m²	2 buildings
3	building density	37%	
4	total number of households	72	24 of 130m² 48 of 90m²
5	volume fraction	2.24	
6	green area coverage	51%	≥30%
7	volume of parking	41%	≥10%

tongli town(From Internet)

[site photo]

[site map]

[general survey of houses]

Cars line

Car Parking

Park
House

Funcational partition

Sidewalk

[site plan]

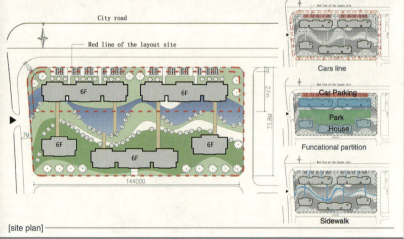

Park in Building Shadow is avoid of sun exposure

Green space to reduce heat island effect

[House]

Light energy house 光动能

单体 Two

[设计说明] Design Notes

考虑到基地自然环境与气候条件，从总体布局到建筑设计注重太阳能的利用，以创新型的可变反射率的围护结构，结合江南水乡的传统建筑造型，能很好地融入当地的自然环境中，创造出和谐、生态的自然环境。

Consider of the natural environment and climatic conditions.from site design to building design all of focusing on use of solar energy and use innovative variable reflectivity of the building envelope, combined with the traditional architectural style, well integrated into the local natural environment, to create a harmonious ecological house.

area: 91.2 m²

area: 129.8 m²

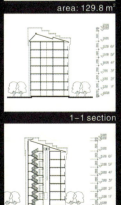
1-1 section
2-2 section

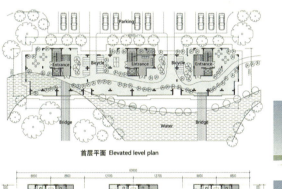
首层平面 Elevated level plan

标准层平面 Standard Floor Plan

北立面 North facade　东立面 East facade
南立面 South facade　西立面 West facade

策略 Three

Light energy house 光动能

2272

[技术说明]
Technology Notes

通过计算机模拟的方式对建筑规划以及单体进行分析,以此得出一个最优的结果。

Through computer simulations for planning, and analysis of resolution in order to arrive at a best result.

[建筑技术]
Building Technology

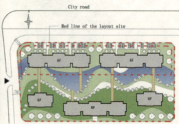

Analysis of outdoor ventilation

 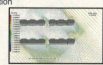

As shown: Compared with the traditional before and after arrangement, "the three in before and two in after" is Benefit for outdoor ventilation.

Analysis of the outdoor light environment.

As shown: Compared with the traditional before and after arrangement, "the three in before and two in after" is benefit for light environment is smaller shadow area.

Great southern sloping of the roof is benefit for arranging solar collector.

Block wind wall is resist northwest winds in winter (red wall).

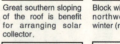

Continuous balcony is a good external shading in summer.

Smaller shape coefficient is good for saving energy.

Use Bernoulli's principles for pulling air in stairwell.

open the door for ventilation in summer

close the door for ventilation in winter

Overhead ground design
Overhead ground is good for ventilation

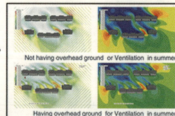

Not having overhead ground or Ventilation in summer

Having overhead ground for Ventilation in summer

Wecan see having overhead ground be good for ventilation in summer.

No door in winter — Close door in winter
No door in winter — Close door in winter

We can see having door be good for resisting wind in winter.

[户型技术]
Unit Technology

[Analysis light environment]
Certificate of analysis by ecotect radiance of light environment is good(outside full cloudy 4500 lx)

[Optimization pipeline layout]
- air pipe
- water pipe

Suit depth design is benefit for both saving-energy and ventilation.

LIGHT energy house 光动能

构造 Four

[设备技术] Equipment Technology

地源热泵制热系数高达3.5~4.5，而锅炉仅为0.7~0.9，可比锅炉节省70%的能源和40%~60%的运行费用，制冷要比普通空调节能30%左右。吴江地区处于夏热冬冷地区，适合使用地源热泵系统。通过提供的气象条件，可以得出吴江地区需要补充地热，利用可变反射率的围护结构，选择性地吸收太阳能来补充地热。同时，由于全年湿度较大，适合用温湿度独立控制空调系统。

Ground source heat pump heating coefficient is up to 3.5~4.5, but boiler is only 0.7~0.9, comparable to the boiler to save 70% energy and 40%~60% of operating costs, refrigeration air conditioning energy than the average 30%. Wujiang in area of hot summer and cold winter zone, suitable for use ground source heat pump system, through the provision of meteorological conditions, can be found the building in Wujiang needs to add to heat. Using the maintenance of the structure variable reflectivity, selective absorption of solar energy to supplement to get hot. The same time, high humidity throughout the year for air conditioning with independent temperature and humidity control system.

智能幕墙 smart wall

By controlling the rotation of shutter to get the different reflectivity wall. for example: white in summer, black in winter.

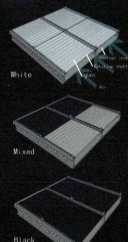

White / Mixed / Black

智能系统 Smart system

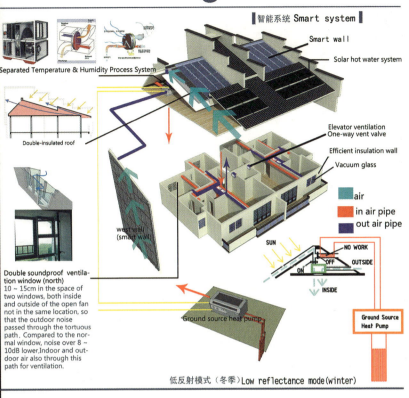

- Smart wall
- Solar hot water system
- Separated Temperature & Humidity Process System
- Double-insulated roof
- Elevator ventilation One-way vent valve
- Efficient insulation wall
- Vacuum glass
- air / in air pipe / out air pipe
- west wall (smart wall)
- Ground source heat pump
- Ground Source Heat Pump

Double soundproof ventilation window (north)
10~15cm in the space of two windows, both inside and outside of the open fan not in the same location, so that the outdoor noise passed through the tortuous path. Compared to the normal window, noise over 8~10dB lower. Indoor and outdoor air also through this path for ventilation.

低反射模式（冬季）Low reflectance mode(winter)

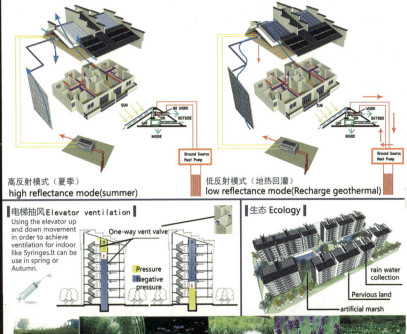

高反射模式（夏季）high reflectance mode(summer)

低反射模式（地热回灌）low reflectance mode(Recharge geothermal)

电梯抽风 Elevator ventilation
Using the elevator up and down movement in order to achieve ventilation for indoor like Syringes. It can be use in spring or Autumn.

- One-way vent valve
- Pressure / Negative pressure

生态 Ecology
- rain water collection
- Pervious land
- artificial marsh

三等奖
Third Prize

项目名称：檐续曦望
　　　　　Continuation Of Hope
作　　者：李　斌、刘清越、吴林娟、
　　　　　李　振、宋　祥、鹿少博
参赛单位：山东建筑大学建筑城规学院

专家点评：

方案较好地利用了传统的围护结构和不同的主被动太阳能利用技术。太阳能板的长短布局与线面组合与墙、柱、窗、檐等建筑元素有机结合，使建筑既具江南水乡韵味，又显绿色建筑技术特征。

The scheme is better to apply traditional inclosing structure and active and passive solar technology. The building has the charm of water village in southern China and also the character of green building technology because of organic combination of solar boards with different length and building elements such as walls, columns, windows and eaves.

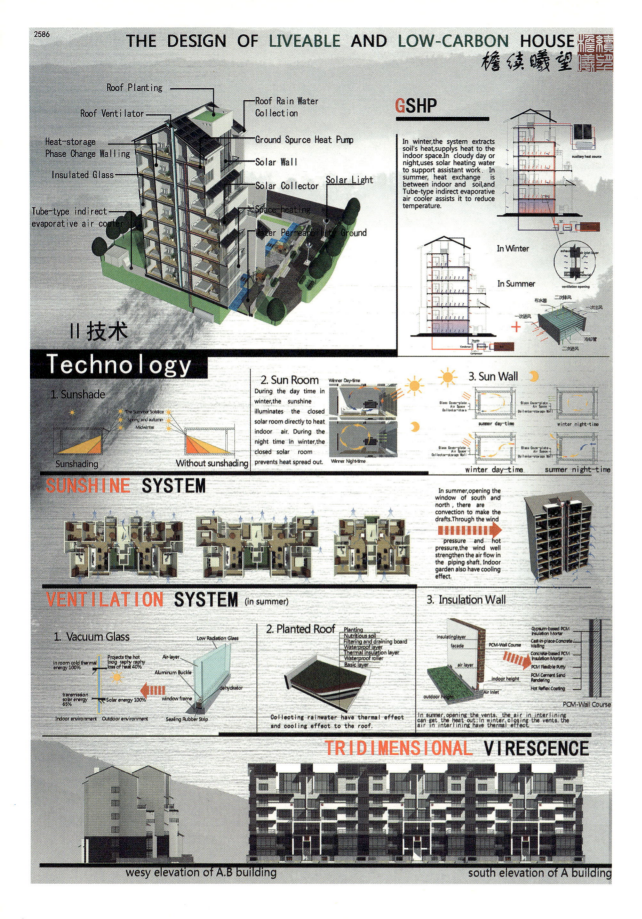

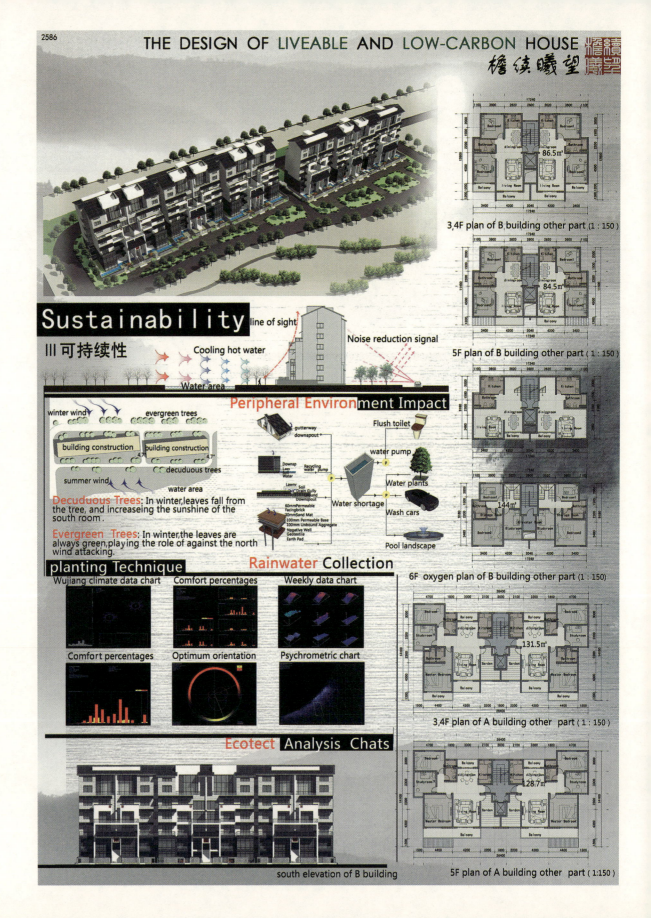

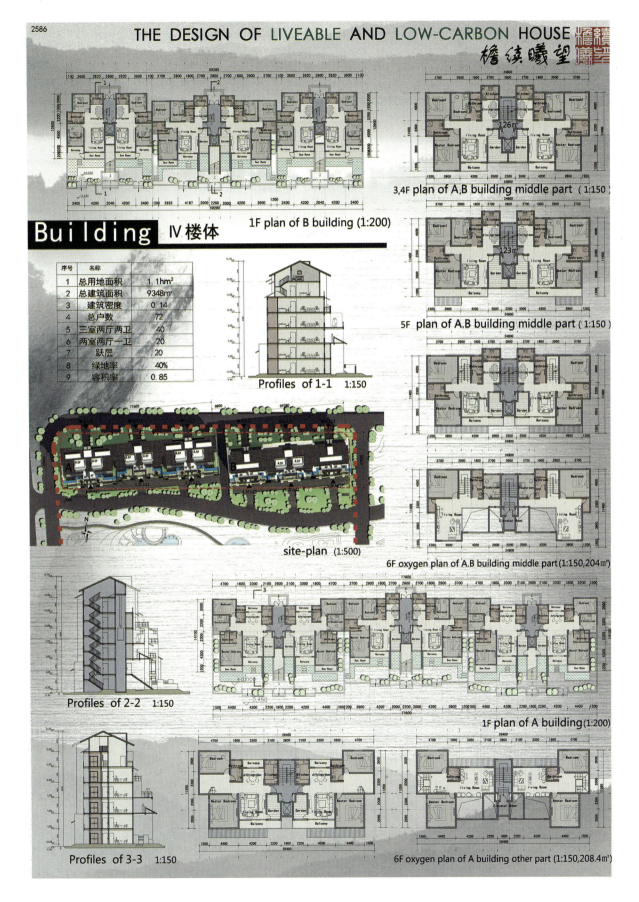

三等奖
Third Prize

项目名称：塞外风光
ECO Residence in Northern China

作　者：徐燊、叶天威、郝铭、郑前、张立名

参赛单位：华中科技大学建筑与城市规划学院

专家点评：

方案较好地融入了西北地域及基地地形特点，规划合理，建筑造型与空间组织较好。住宅标准层结合拔风井形成私密、有层次的内部空间。包括雨水收集和风力发电在内的屋面和南立面设计较好地展示了主被动太阳能技术的建筑应用。

The scheme is better to integrate the character of northwest area and construction site. The planning is rational and it is better on building form and space organization, too. On standard storey a private, layered space is formed combined with ventilation duct. The roof installed with rain water collector and wind power generation and the design of south facade show better application of active and passive solar technology.

塞外风・光
ECO RESIDENCE IN NORTHERN CHINA

DESIGN DESCRIPTION

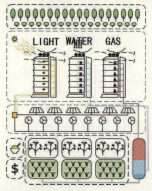

本设计多方面考虑呼和浩特当地建筑特色和气候因素，在降低建筑自身能耗的基础上充分利用太阳能、风能等可再生能源，营造舒适的人居环境。

场地设计： 种植本土农作物、风车。生态停车场太阳能景观小品。

建筑节能： 南向房间设置阳光房和Trombe Wall屋顶设置光伏/风发电系统，楼梯间成为通风核，将增加湿度的空气带入室内。楼梯间立体绿化提供私密交流空间。

宜居住宅： 住宅采用创新的体系；户型多样化设计；底层和屋顶设置阳光间，在寒冷的冬季提供公共活动空间。

SITE PLAN 1:800

Solar panels　Vegetation　Wind turbine　Farmland
Roof　60m²　90m²　Bottom floor
South facade　Ventilation core　North facade　West facade

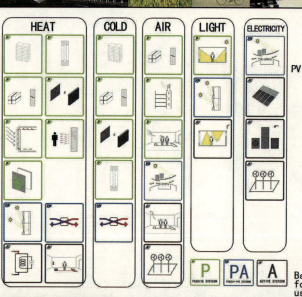

HEAT　COLD　AIR　LIGHT　ELECTRICITY

P PASSIVE SYSTEM　PA PASSIVE-ACTIVE SYSTEM　A ACTIVE SYSTEM

ENERGY THEMES & SITE SECTION

Solar house, Green roof, PV/wind hybrid sail, PV pavilion, green noise barriers, Eco parkinglot, Low-noise wind turbine, Greenhouse, Eatable landscape

Basic plan for multiple use　Module block　Open colunms and floors　Public spaces　Expandable units

ECONOMIC INDICATOR

Prosite area: 11000m²
Building area: 22450m²
Floor area ratio: 2.04
Building density: 16.7%
Greening rate: 53.4%
Number of floor: 11
Parking: 22
Number of Households:190
（90m²：90　60m²：100）

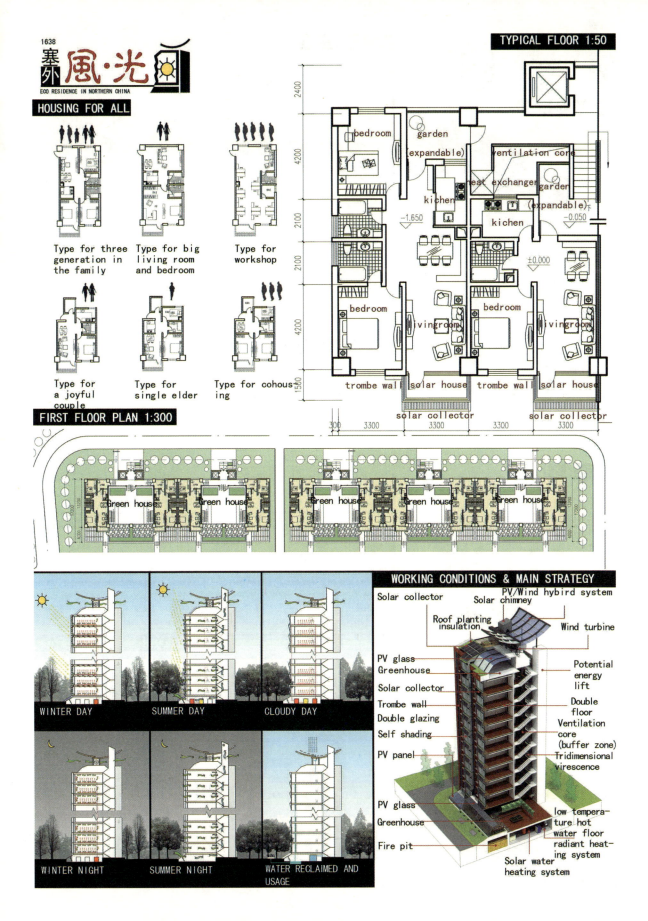

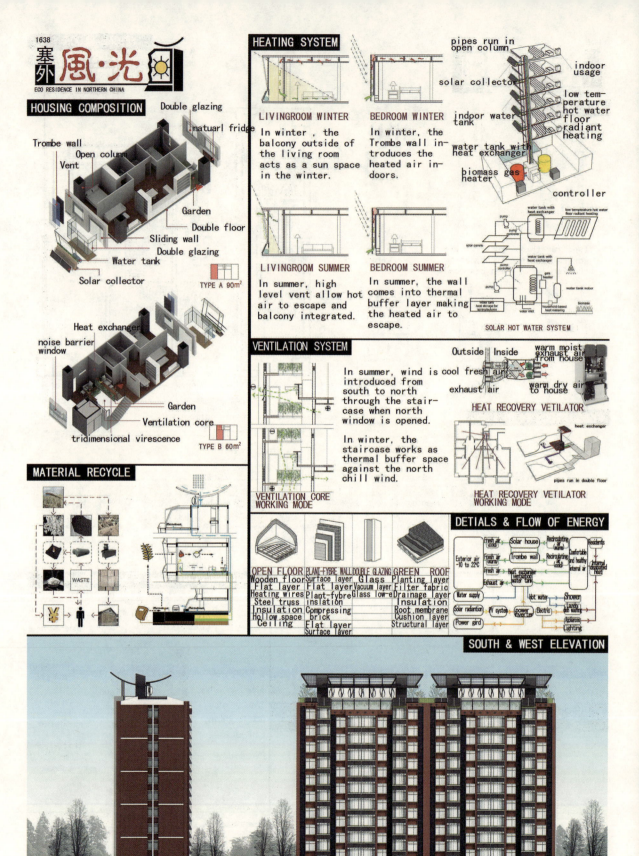

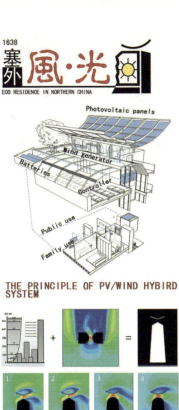
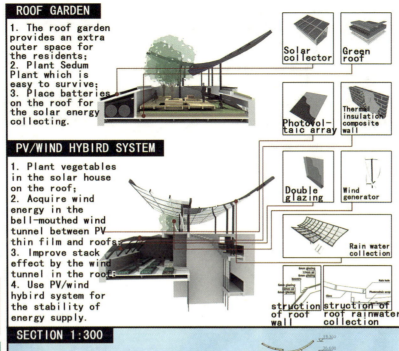
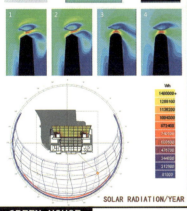
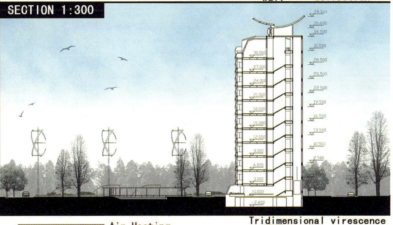
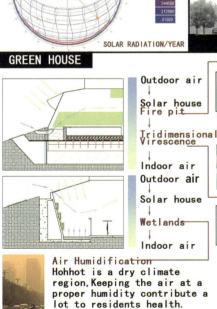
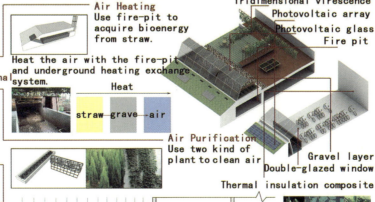

Lakeside in the Sunshine 阳光下的湖畔 1

code: 1153

优秀奖
Honorable Mention Prize

项目名称：阳光下的湖畔
Lakeside in the Sunshine

作　者：张方晴、陶如、马全明、
张一兵、田海鹏

参赛单位：中国矿业大学力学与
建筑工程学院建筑学系

Local Characteristic
- crop
- reed
- tree
- building
- street
- residence

Introduction
Background
Location

China　Suzhou　Wujiang　Tongli

Design Description

North of the base is residence, south is landscape. Program from planning to single to node is compliance ecological principles that with solar energy, reducing energy consumption and reduce pollution. Underlying of building using water and plant to regulate micro-climate and reduce noise. It is with a unique form of canal towns south of Yangtze, and using steel, soil, reed block which are recyclable materials. In passive solar energy utilization, it uses pull air shaft, sun room, tridimensional virescence, plant heat insulating, attic ventilation insulating, adjustable external shading, ventilation and sound insulation window. It also has a special design in hot and cold parts of the doors and windows outside. In active solar energy utilization, it mainly uses solar plus supplementary system, solar panels and ground source heat pump. In landscape design, using trees and lakes in the base, and solar lights and solar battery charging station besides, deeper interpreting the theme.

For local residents
Knowing of the solar building: know / a little / don't know / never heard

Heating way in winter: others / stove / warm gas / air condition / none

Cooling way in summer: others / fan / air condition / none

Research

Environment
The base is located in Wujiang, Suzhou, Jiangsu Province (cold winter and hot summer), with the north of historical and cultural city "Tongli town" and the west of Tongli Lake. The base is in the south of urban trunk Jiangxing dong Road. Planned land area is 11,000 ㎡, and the implementated area is 3890 ㎡. The project is positioned as a high-end solar energy residential community.

90㎡ Unit

130㎡ Unit

设计说明

基地北侧规划住宅，南侧景观。方案从规划到单体到节点都充分遵循利用太阳能，降低能耗，减少污染的一体化生态设计原则。建筑底层利用水源和种植调节建筑微气候及降噪。形式上采用了江南水乡的建筑特色，还使用了钢、土、芦苇砌块等当地可回收的材料。被动式太阳能设计采用了拔风井、阳光房、立体绿化、植物隔热、阁楼通风隔热、可调节外遮阳、通风隔声窗等手段，外门窗等冷热桥部位也进行了特别的设计。主动式太阳能技术采用太阳能辅助加热系统、太阳能电池板以及地源热泵。景观设计中，利用基地树种和湖水，还设计了太阳能路灯和太阳能充电站，深层次诠释了主题。

Economic and Technical Indicators

Total land area: 11000㎡
Total floor area: 8946㎡
Building density: 12.5%
Total households: 60
FAR: 0.81
Green rate: 54%
Ground parking rate: 115%

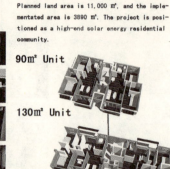

Lakeside in the Sunshine 阳光下的湖畔 2
code: 1153

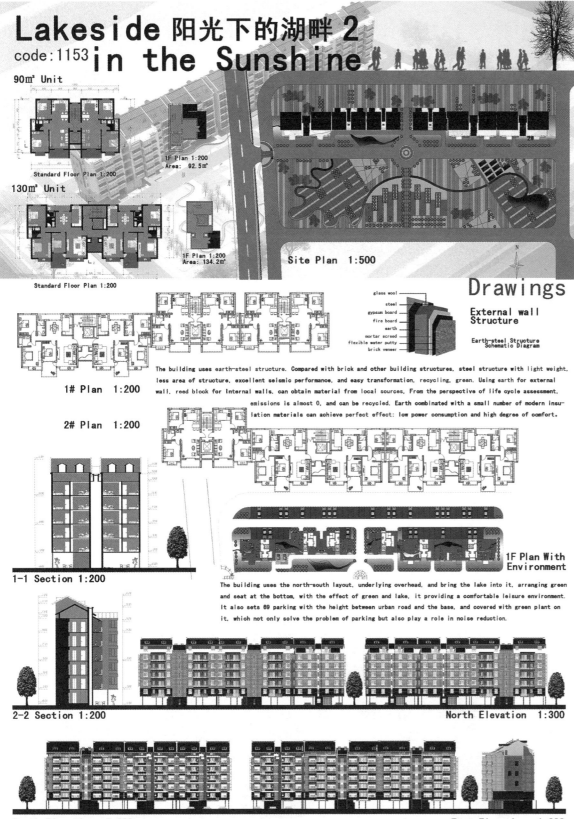

The building uses earth-steel structure. Compared with brick and other building structures, steel structure with light weight, less area of structure, excellent seismic performance, and easy transformation, recycling, green. Using earth for external wall, reed block for internal walls, can obtain material from local sources. From the perspective of life cycle assessment, emissions is almost 0, and can be recycled. Earth combinated with a small number of modern insulation materials can achieve perfect effect: low power consumption and high degree of comfort.

The building uses the north-south layout, underlying overhead, and bring the lake into it, arranging green and seat at the bottom, with the effect of green and lake, it providing a comfortable leisure environment. It also sets 69 parking with the height between urban road and the base, and covered with green plant on it, which not only solve the problem of parking but also play a role in noise reduction.

Lakeside in the Sunshine 3
code: 1153
Technology

sets 69 parking with the height between urban road and the base and covered with green plant on it, which play a role in noise reduction.

Technology confluence

- Tridimensional virescence
- Solar chimney
- ventilation window
- Roof insulation
- Earth-steel structure
- Solar cell
- Solar chimney
- Reed block
- Solar collector
- Ventilation and sound insulation window
- Solar cell
- Heat storage floor
- Tridimensional virescence
- Sun room
- Green wall
- Tridimensional virescence
- Sunshading
- external windows insulation system
- Solar parking lot
- Double low-e glass
- Water cooling
- Solar lights
- Overhead ground
- Ground Source Heat Pump

Solar cells — Light — Controller — Battery

Thermostat — Temperature probe
- Ground floor
- Concrete
- Composite insulation
- Structural layer
- Ground floor
- Heating
- Composite insulation
- Connection frame
- Aluminum foil

water collect

Double low-e glass

Adjustable external shading system

change the angle of shutter   change the length of roller shutter

Interior case with shutters open

Interior case with shutters closed

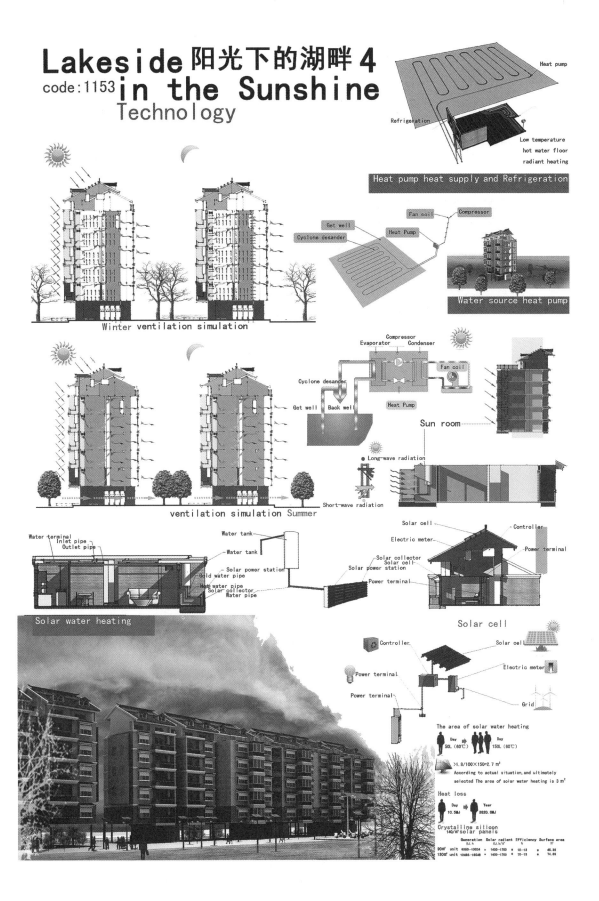

优秀奖
Honorable Mention Prize

项目名称：寻光之旅
In Search of the Way of Sunlight

作　者：张瑞娜、刘　鸣、范　悦、
　　　　袁　杰、李　莉、陈　滨、
　　　　张宝刚、段苏桐

参赛单位：大连理工大学建筑与艺术学院

寻光之旅 IN SEARCH OF THE WAY OF SUNLIGHT

There are many soulful stories about sunlight, such as Yi, a strong young man shooting nine suns and Kuafu, a persistent man, being bent on chasing the sun lighting. Today we become persons who are bent to on exporing solar technology. We have different purposes but the same spirit and counge with them, so we believe that solar technology can bring energy, health, and comfortable to our life.

the Archer Houyi Shooting the Sun   the Giant Kuafu Chasing the Sun

INSOLATION PROSESS ANALYSIS

We are walking in search of the Way of SunLight…

SITE PLAN 1:500

orginal

adjust

Optimum

BEST ORIGINAL

SOLAR RADITION

设计说明：

技术改变生活，生活推动技术。本设计方案结合新颖、实用的主被动技术方法，营造资源循环、生态健康的住宅及其环境。本设计主要特色之处如下：独立的阳光间；曲面集热墙；自然通风器；绿化种植；太阳能毛细管网结合供暖；景观式防噪声屏障；雕花窗式景观墙；太阳能LED照明系统；雨水收集；厨房及卫生间垃圾粉碎集中处理。

Design focus:

Aims of the plan are to create the resource-recycling, ecological and health of the resident environment by the application of active and passive technologies. Technology has changed our life, which life promote technologies. Main features of the design: a separate sun room, Curve surface hot collector wall, natural ventilator, Planting, capillary network with solar heating, landscape-style antinoise barrier, solar LED lighting systems, rainwater collection, concentrated disposal of crushed kitchen and bathroom garbage.

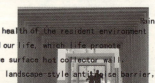

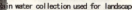
Rain water collection used for landscape

Ventilation combined with landscape

parking lots combined with landscape

solar LED lighting systems with landscape

noiseproof combined with landscape

entrance of underground ventilation

PERSPECTIVE

寻光之旅  IN SEARCH OF THE WAY OF SUNLIGHT

Index

Planning Area	11000hm²
Building Area	13440m²
Floor Area Ratio	1.22
Building Density	0.204%
Green Rate	60.9%
Total Number Of The Program	60 households
Total Number Of Planning	96 households
Ground Parking Rate	11.5%

130m² Analysis Of Energy Saving

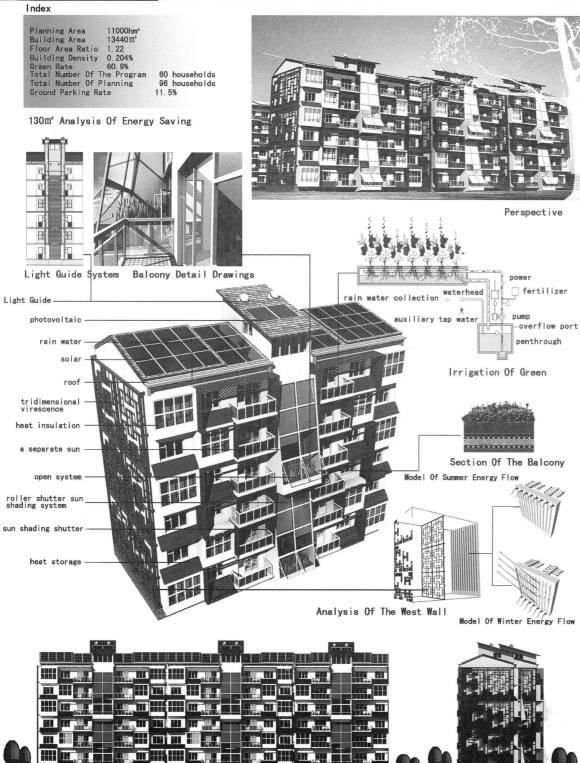

Light Guide System   Balcony Detail Drawings

- Light Guide
- photovoltaic
- rain water
- solar
- roof
- tridimensional virescence
- heat insulation
- a separate sun
- open system
- roller shutter sun shading system
- sun shading shutter
- heat storage

Perspective

Irrigation Of Green

Section Of The Balcony

Model Of Summer Energy Flow

Analysis Of The West Wall

Model Of Winter Energy Flow

South Elevation 1:200   West Elevation 1:200

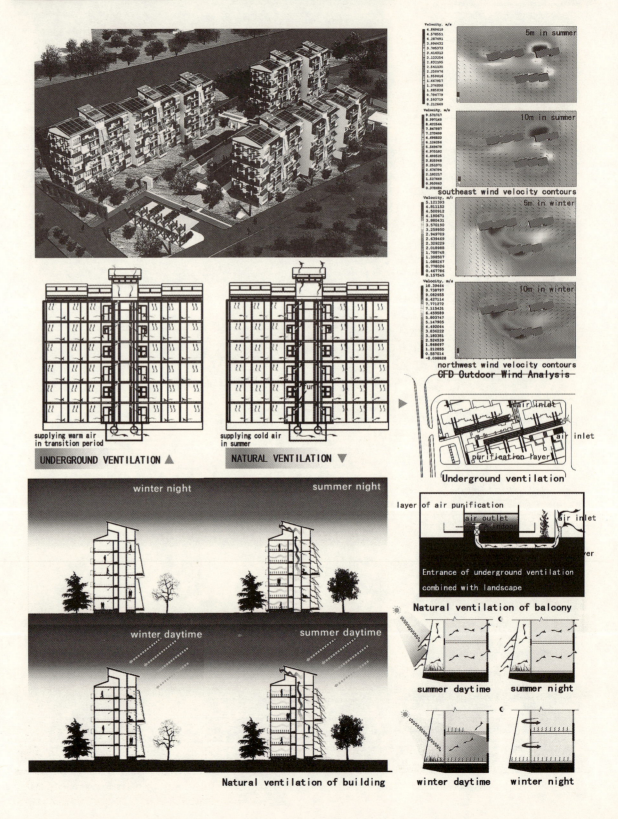

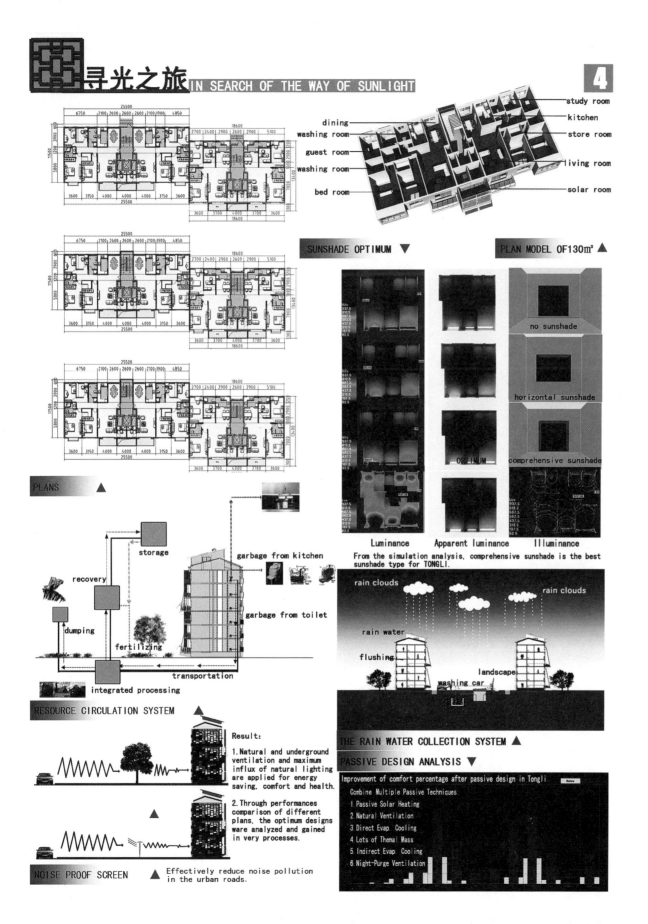

优秀奖
Honorable Mention Prize

项目名称：太阳能花园
　　　　　Solar Garden
作　　者：李 欣、江 雯、杨维菊
参赛单位：东南大学建筑学院

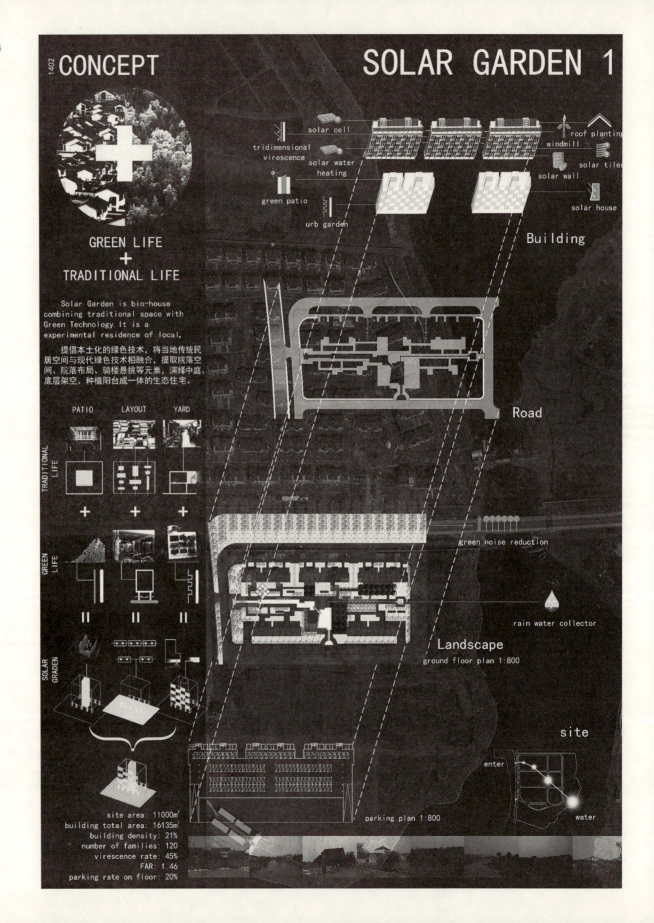

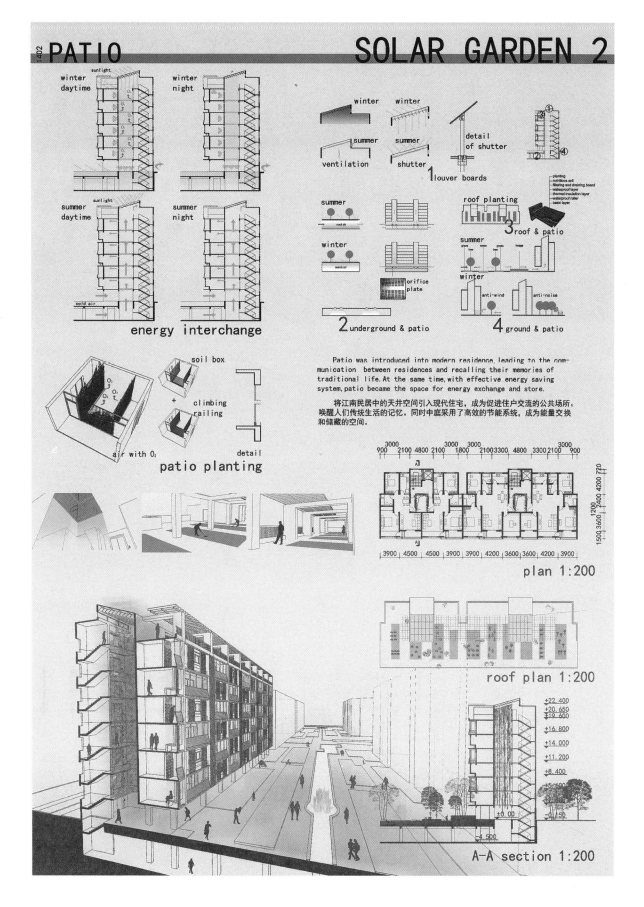

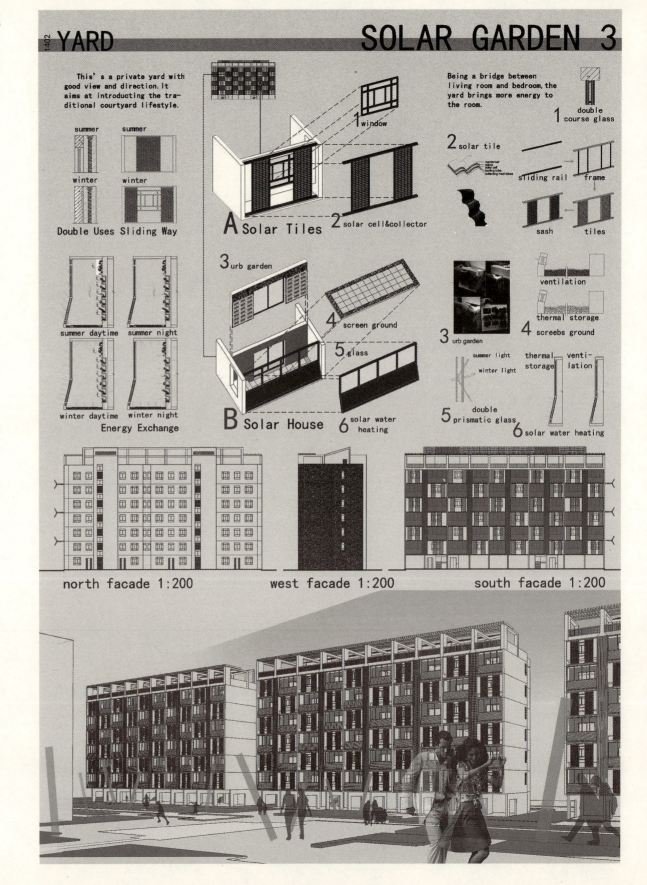

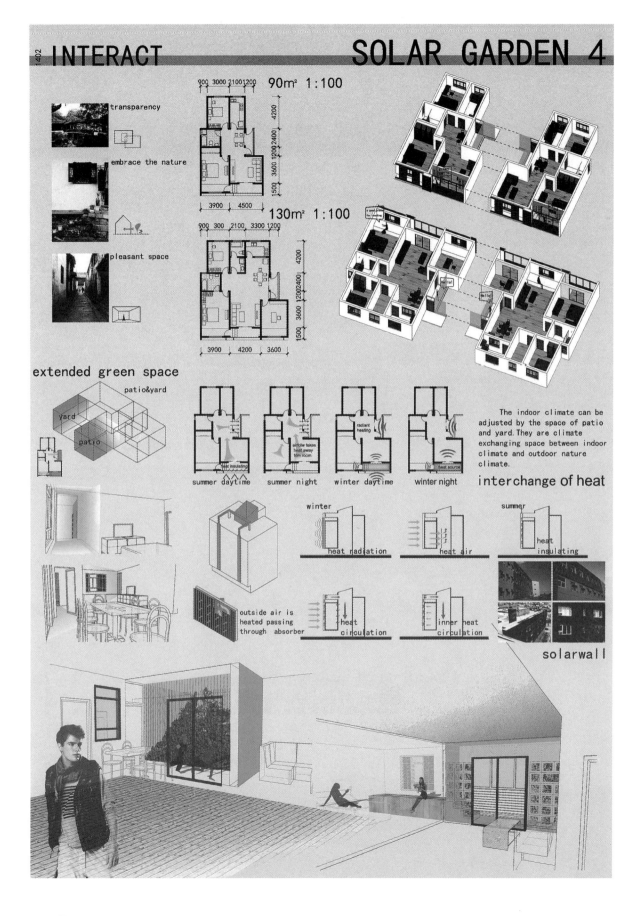

优秀奖
Honorable Mention Prize

项目名称：富土阳光
　　　　　Solar Energy Residence in Tongli

作　　者：陈　曦、倪江涛、王辰君、
　　　　　王润栋、郑康奕、刘裕辰

参赛单位：上海大学建筑系

富土陽光

低碳·阳光·回归 ——吴江市低碳宜居住宅设计

SOLAR ENERGY RESIDENCE IN TONGLI

What can we make full use of the design?

- Temperature
- Solar Radiation
- Wind Speed
- HDD
- Humidity
- CDD

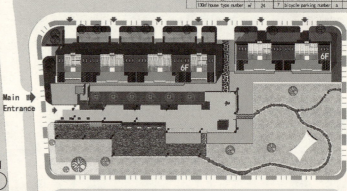

Main Entrance

Master Plan 1:500

No.	Name	Unit	Quantity	No.	Name	Unit	Quantity
1	construction site area	m²	4500	3	greed space area	m²	3600
2	total floor space	m²	9000	4	shape coefficient		0.32
	pavement space	m²	630	5	greening rate	%	32.7%
	90m² house type number		12	6	floor area ratio		0.82
	130m² house type number		24	7	bicycle parking number	a	75

设计说明：

江苏吴江地区低碳宜居住宅设计，考虑基地自然环境与气候条件，从整体布局到建筑设计注重以太阳能、风能为主的自然资源的结合与利用。采用高科技、高效率策略，作一体化的节能设计。该设计体现宜居住宅的概念，将阳光、绿化、自然引入室内，同时又能感受到生态技术带来的舒适。

The design of low-carbon residents in Jiangsu Wujiang region, is found upon the analysis of actual situation (nature, environment, climate, condition), focusing on making use of solar energy and wind energy source. We make the integration of energy-saving design by using high-tech, high-efficiency strategy. The design embodies the concept of livable housing, introducing the sun, green and natural to the interior, While making you feel the comfort of Ecological Technology.

Tongli Town, south of the six famous rivers and lakes, is a typical water village with a long history and style, surrounded by five lakes, the town by the river network is divided into seven islands. And the design bases to the same Lake is located in one of the West Bank, north side of town.

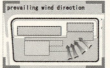 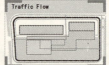 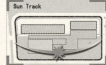
prevailing wind direction | Traffic Flow | Sun Track

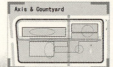 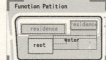 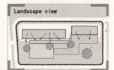
Axis & Countyard | Function Patition | Landscape view

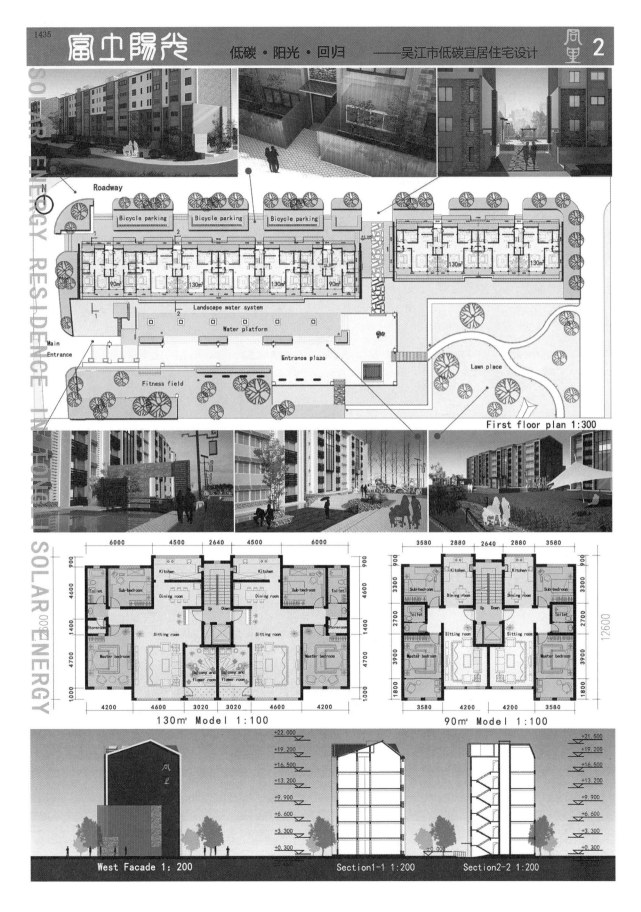

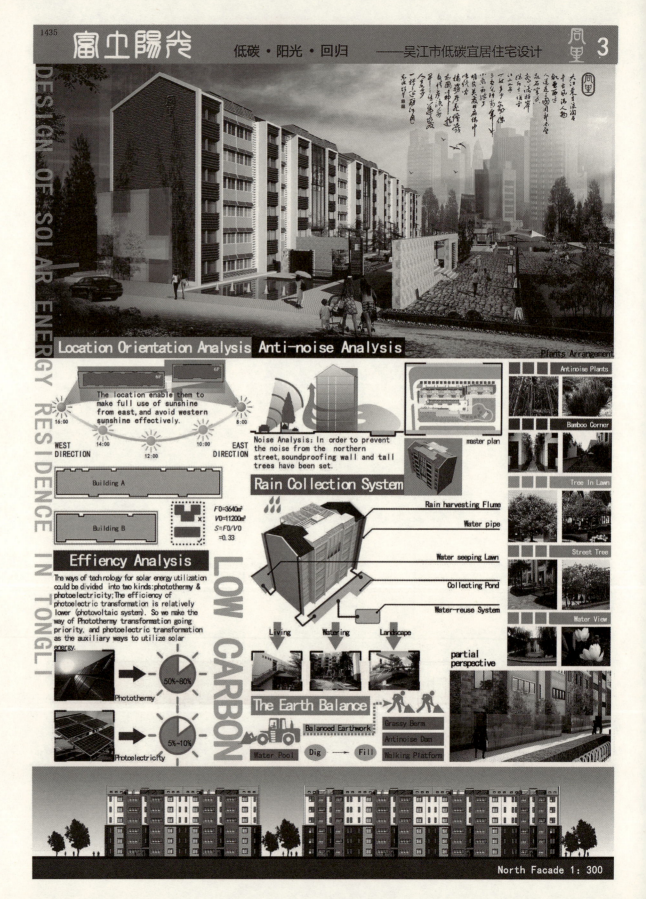

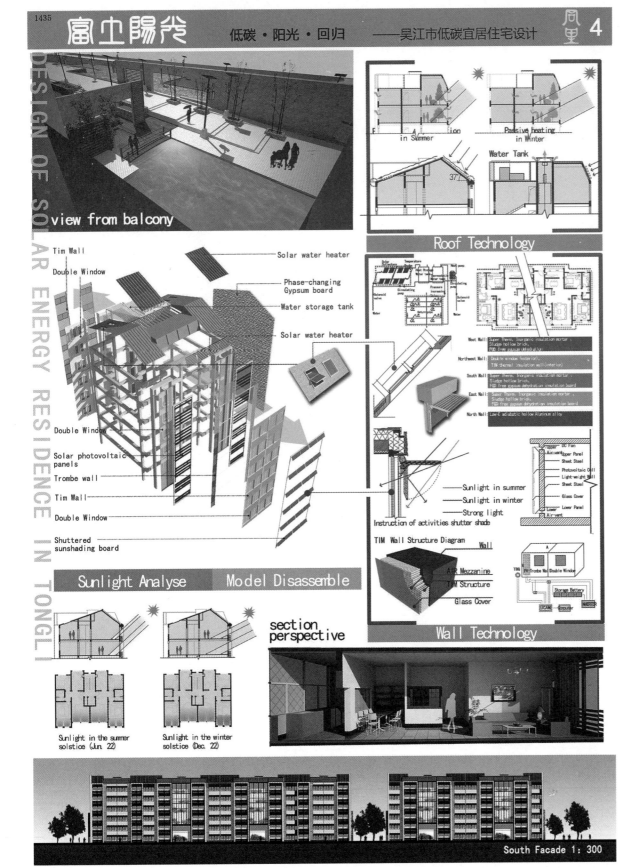

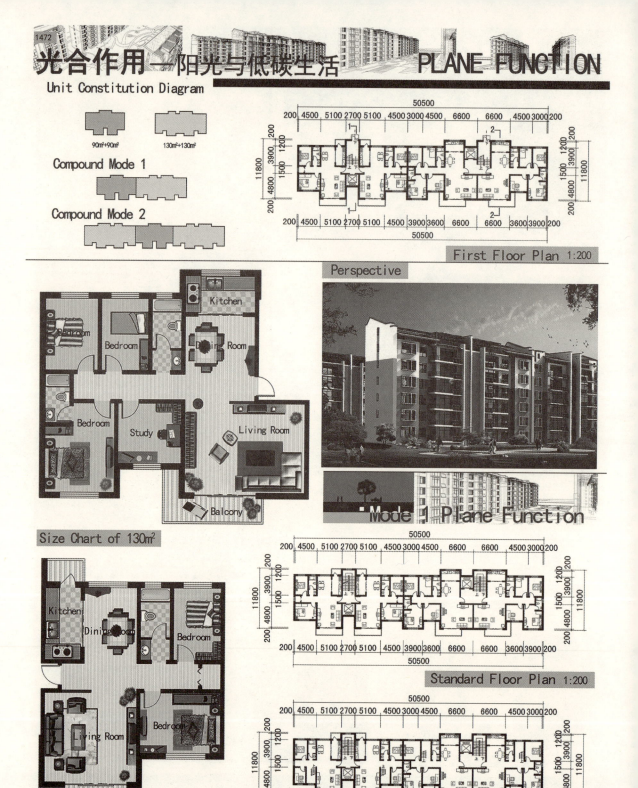

光合作用—阳光与低碳生活

PLANE FUNCTION

Local Perspective

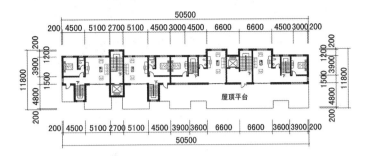
Spring Layer Plan 1:200

1-1 Section 1:150

Perspective

Mode 1. Plane Function

2-2 Section 1:150

South Elevation East Elevation

LOW-CARBON LIFE WITH SUNSHINE

TECHNOLOGIES

光合作用—阳光与低碳生活

Building Lighting Diagram

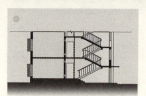

Summer sun elevation angle greater level of shading can effectively block solar radiation.

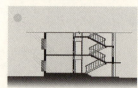

Winter sun elevation angle is small, light can enter the room.

Building Natural Ventilation Diagram

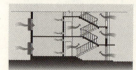

Set indoor wall with vents and stairwells, to strengthen the chimney effect.

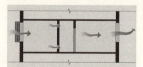

Set in the indoor wall vents, to enhance natural ventilation in buildings.

Building pressure ventiltion diagram.

Ventilation and moisture-proof bottom of overhead.

Schematic diagram of the stairwell ventilation stack effect

Feedback Elevator

Pulling Lead-free installation gear at the top can save building space enormously.

Using "under the lead-style" drag index structure saves a certain space on top.

The elevator control cabinet is designed more "slim", which can be set inside the shaft.

When the weight the counter balance of is greater than the car side, the counter balance streams down by the gravity and drive permanent magnet electric rotating rotor, the elevator produces the electrical power.

Low-temperature Hot Water Radiant Floor

Indoor Temperature Distribution Table

the comfortable feeling / air temperature heating

radiant floor heating / radiator heating

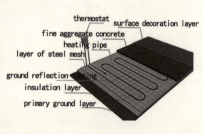

Schematic Diagram of Low-temperature Radiant Floor
- thermostat
- surface decoration layer
- fine aggregate concrete
- heating pipe
- layer of steel mesh
- ground reflection
- insulation layer
- primary ground layer

- planting soil
- non-woven filter layer closure
- QIFENG
- composite insulation board
- structure layer
- roof waterproofing layer

Roof Garden: increase of the rate of virescence and reduce the impact of solar radiation on the room, heat island effect and the visual pollution and improve the vision of the rate of green. Urban green roof is an effective way to solve the lack of public green spaces in the city center and improve eco-green performance.

Roof planting

Solar Photovoltaic

building maintenance
Photovoltaic panels
LED
DC current regulator
DC power regulator
AC
220V AC

The roof garden of the glass top: amorphous silicon thin film solar panels, which can not only absorb solar radiation power generation, but also have a certain transparency to ensure that the amount of light for plant growth.

Comfort Zones

Comfort Level

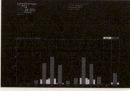

This design apply several low-carbon energy conservation measures such as sunshade, natural ventilation, low-temperature radiation floor, solar energy photovoltaic technology and so on. Through the Ecotect software simulation, we can see from the chart that the residential comfort zones (line diagram) and human comfort level (bar chart) have a great improvement. Through applying these measures, we not only save lots of energy, but also make a comfort thermal environment for users.

LOW-CARBON LIFE WITH SUNSHINE

光合作用—阳光与低碳生活

PLANNING
Design Introduction

This design mainly solve contemporary problems of low-carbon energy of residential buildings, and the design units use small depth and can solve the problem of ventilation and lighting. In the case of construction cost not increasing substantially, we can short the heating period of winter and realize the ventilation in the summer breeze or even under the conditions of no wind by passive solar energy technologies, active solar energy technologies as well as some building components technology. Through the integration of these technologies and building functions, building construction, we can achieve the integration of the energy conservation design and architectural design.

Aerial View

设计说明：

本设计主要从解决当代集合住宅建筑的低碳节能问题入手，设计户型采用小进深，能够很好地解决通风采光问题。通过被动式太阳能技术、主动式太阳能技术以及一些建筑构件节能技术的应用，在不大幅增加建筑成本的情况下，冬季可缩短采暖期，夏季可在晴天微风、甚至无风条件下实现通风。通过这些技术与建筑功能、建筑构造的整合，达到节能设计与建筑设计一体化。

General Plan 1:500

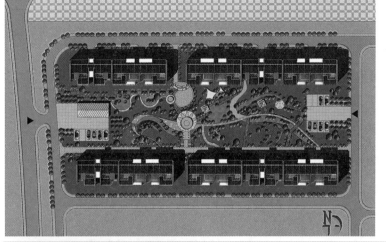

Economic and technological indexes

1	floor area	1.1 hm²
2	building area	18467 m²
3	building density	25%
4	households	120
5	floor area ratio	1.7
6	greening rate	53%
7	parking lots rate	13.3%

Energy Conservation Strategy

Solar Lights Principle

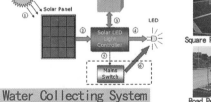

Pavement Materials

Square Permeable Pavement | Parking Grass Brick

Road Permeable Pavement | Road Hollow Concrete

Water Collecting System

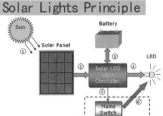

Irrigate | Flushing | Washing

Improve the base and the permeability: we should make the ground water infiltrate into the ground and increase groundwater.

Rainwater collected debr can be used as water after simple treatment.

Plants

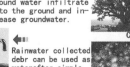
Camphor | Podocarpus

Acacia | Hongyeli

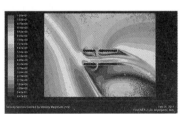
The Wind Simulation

THE URBAN ANALYSIS

The Best Orientation | Monthly Diurnal Averages

 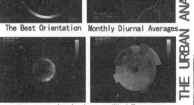
Wind Frequency (Hrs) | Average Wind Temperatures

LOW-CARBON LIFE WITH SUNSHINE

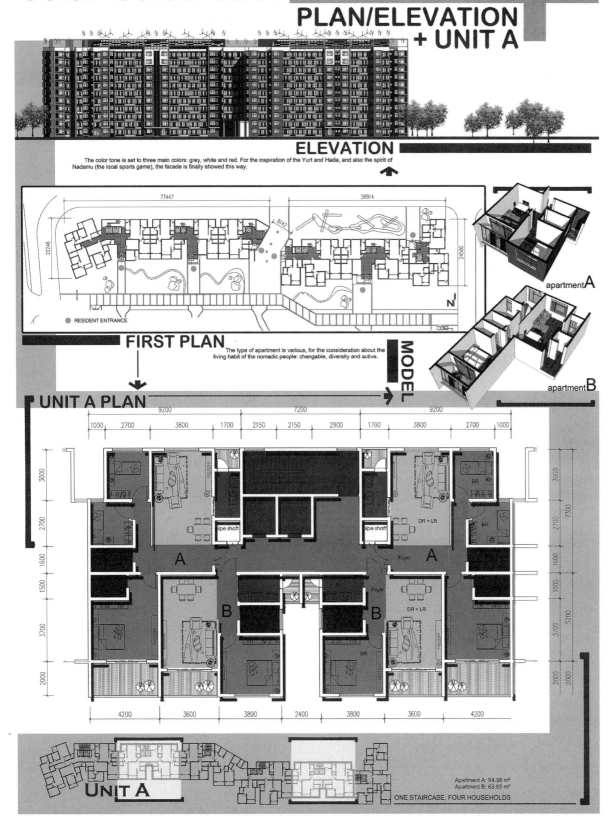

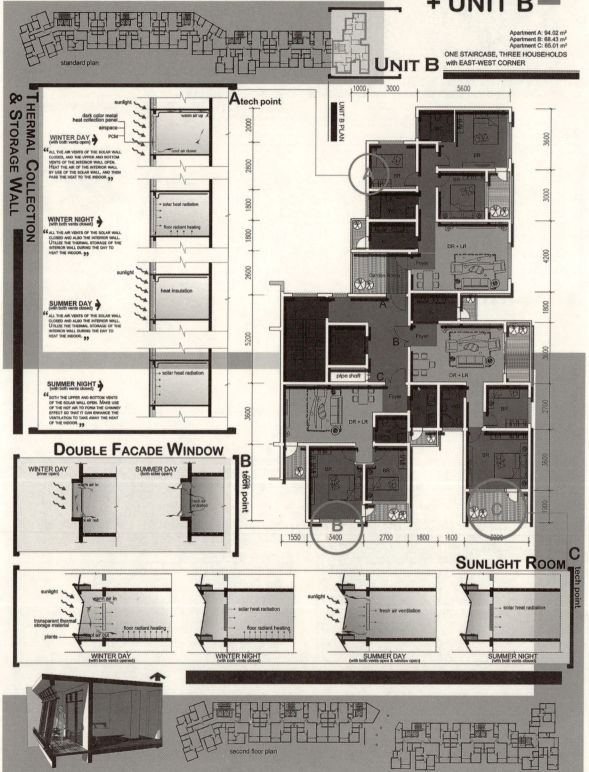

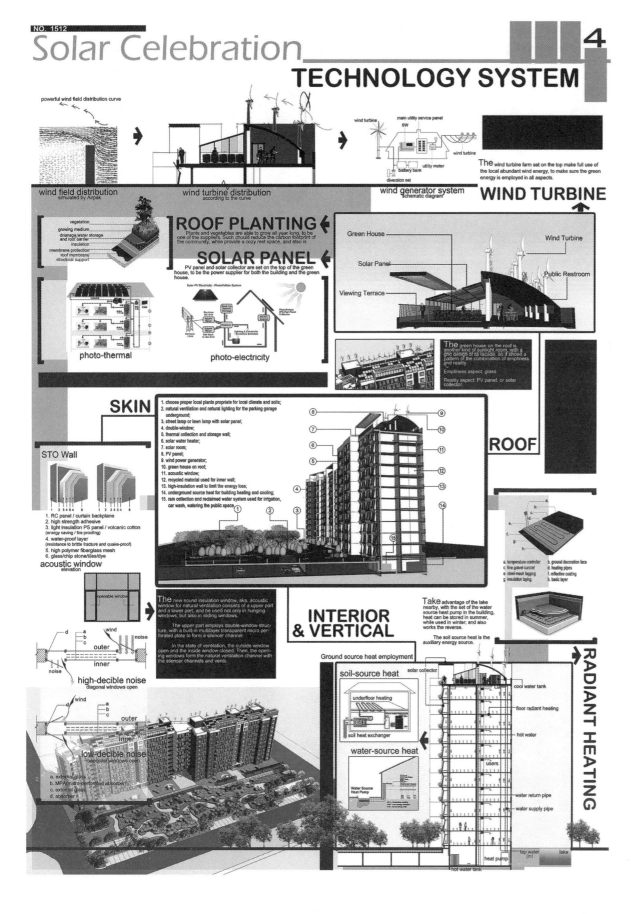

优秀奖
Honorable Mention Prize

项目名称：光之舞
　　　　　Solar-House

作　　者：谢永平、许　杰、孙飞龙、
　　　　　邵云清、王修水

参赛单位：浙江树人大学城建学院

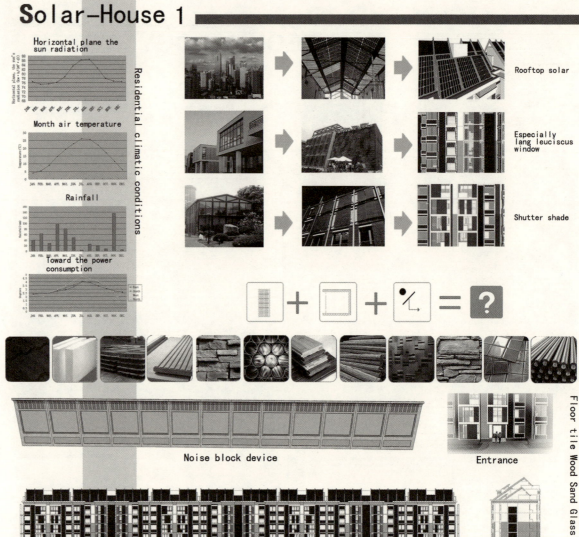

Solar-House 2

Design descriptions

This design depends on the advanced science and technology with innovative thinking, fully utilize solar energy and other natural resources. It makes future living unit more energy-efficient, green.

This design is the theme of the solarhouse. In living unit is designed in the sun room, roof photoelectric, shutter shade ventilated.

With innovative thinking, building do STH unconventional or unorthodox, independent absorb solar, wind. Attain low carbon buildings, livable architecture

Economic and technological indexes

NO.	project	unit	note
1	Total land area	11000m²	
2	Construction area	9688.8m²	
3	Building density	15.3%	
4	Total household	84	
5	Cubage	0.88	
6	Greening rate	32.9%	≥30%
7	Ground parking rate	17.8%	≥10%

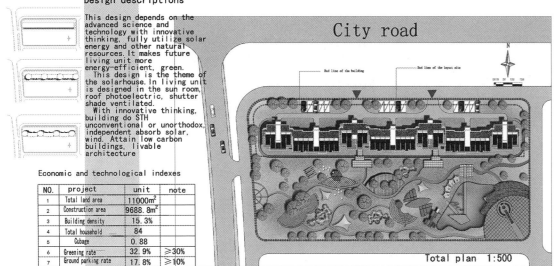

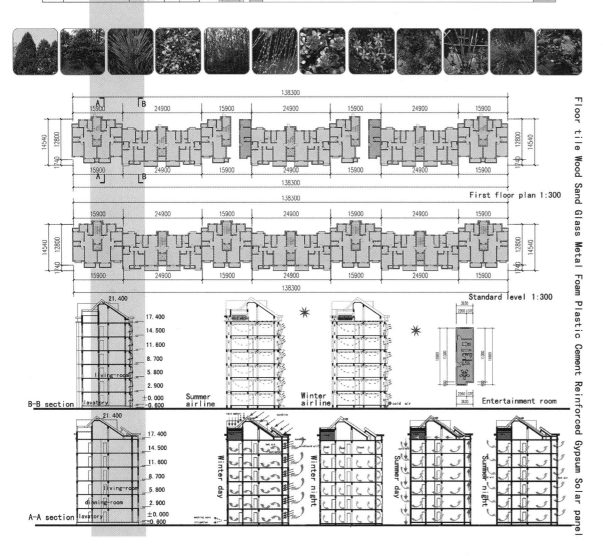

Solar-House 3

1593

Plan employed (130m²)

The analysis of changable balcony1

- Closed
- Sunshine
- AM 8:00~11:00
- 11:00~13:00
- PM 13:00~15:00
- Summer night
- Winter night
- Windows
- Windows closed

1. Changable window
2. Especially lang leuciscus window
3. Changable balcony1
4. Solar collector
5. Solar chimney
6. Ventilation shafts

The analysis of roof

- Collector
- YanGou
- Drain
- Insulation
- Shutter
- Planting
- Nutritious
- Filtering board
- Waterproof layer
- Thermal insulation layer
- Waterproof roller
- Structure
- Roof planting
- Winter Air streamline
- Summer Air streamline

The analysis of Especially lang leuciscus window

- Ourdoor air into
- Solar panels absorb solar energy
- South sun radiation heating
- Outdoor air into
- Solar panels aborb solar energy
- Air preheater

SpinachOnionMelonTomatoesSweetChiliCucumberPotatoCowpeaPumpkinEggplantGreenStrawb

Solar-House 4

Plan employed (90m²)

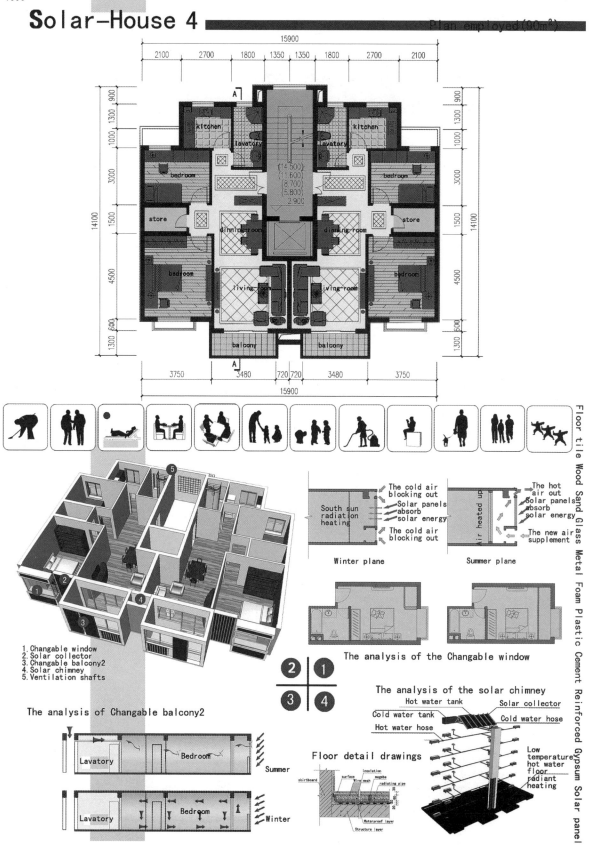

1. Changable window
2. Solar collector
3. Changable balcony2
4. Solar chimney
5. Ventilation shafts

Winter plane / Summer plane

The analysis of the Changable window

The analysis of the solar chimney

The analysis of Changable balcony2

Floor detail drawings

优秀奖
Honorable Mention Prize

项目名称：阳光水榭
　　　　　Sunny Waterside
作　　者：李 亮、陶 然、康玉东、
　　　　　刘杰民、马淑洁
参赛单位：山东建筑大学建筑城规学院

阳光水榭
REGISTER NUMBER: 1607

LIVEABLE AND LOW-CARBON RESIDENCE IN WUJIANG

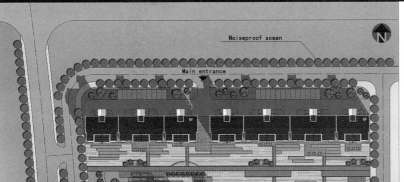

设计说明

临水而居，天井院落是江南水乡的典型居住形式，本案在深入研究当地民居特点后，利用景观水面、底层架空、入户花园等方式营造出一种阳光水榭的乡居意境。水面、院落的合理布局，不仅增添了空间的趣味性，同时改变了建筑周边的小气候，起到节能、宜居的作用。

对阳光、风的引导利用，不失为本案的又一特色。炎炎夏日，加强自然通风的组织，既可以降低室内温湿度，又可改善空气质量，有效地降低空调能耗。寒冷的冬季，新型太阳能技术、地热技术的合理搭配，为悠然的乡居生活平添一丝春色，使绿色、节能的概念成为生活的主题。

Living near the river, the patio courtyard is a typical living form of water town in jiangnan, This program in-depth study of the local residential areas, the use of landscape water, overhead bottom, garden home, etc. to create a sunny waterside of a rural mood. Water and the compound of the rational distribution of space not only adds interest, but also changes the micro-climate around the building, makes the role of livable and energy-saving play.

The use of leading the sunshine and wind, is another feature of the case. In sorching summer, enhancing the organization of natural ventilation, not only can reduce the indoor temperature and humidity, but also improve quality of air, effectively reduces the energy consumption of air conditioning. In cold winter, the reasonable collocation of the new solar technology and geothermal technology, adds a hint of spring for the leisurely country life, makes the concept of the green, energy-saving to become a life theme.

Economic and technical indicators

Number	Name	Units	Amount
1	The total land area	m²	11000
2	The total construction area	m²	8940.6
3	The total family number		84
4	Building density	%	8.85
5	Plot ratio		0.813
6	Greening ratio	%	51.8
7	The ratio of car parking	%	27

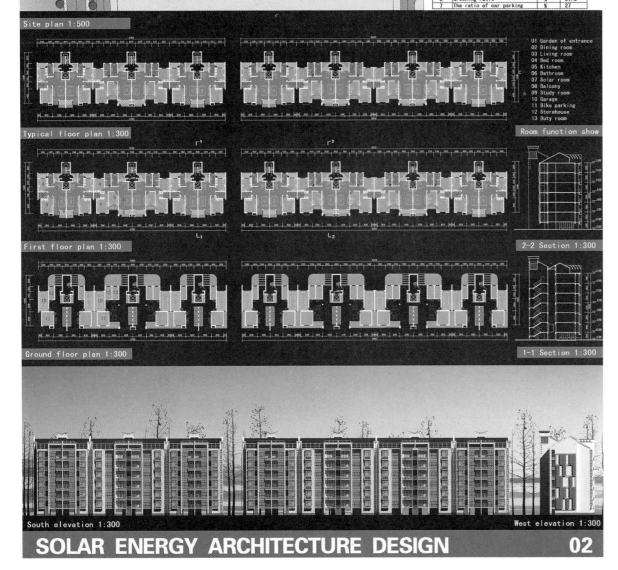

SOLAR ENERGY ARCHITECTURE DESIGN   02

阳光水榭 — LIVEABLE AND LOW-CARBON RESIDENCE IN WUJIANG

REGISTER NUMBER: 1607

Solar—geothermal energy analysis

The central hot water system of the Solar—geothermal energy is a hot water system form of dependable performance, environmental protection and energy saving. the system only uses solar energy and less electricity without any pollution toward the environment. In summer, when solar energy is low, the system will automatically change to the running state of geothermal energy, through the ground—source heat pump that the low temperature energy is obtained directly from the soil, in the double-effect evaporator the refrigeration and air conditioning are implement; in winter, the solar—cooling system is changed to the heat pump cycle way for heating the indoor, when the solar energy is low, the system starts automatically the equipment of the geothermal heat pump, alternatives to heat the indoor instead of the solar energy.

Technology confluence

Labels: Stairwell ventilation; Solar cell; Collector array; Solar collector; Sunshading shutter; Air collector; Solar house; Heat insulation window; Planting roof; Artificial marsh effect; Rain water collection

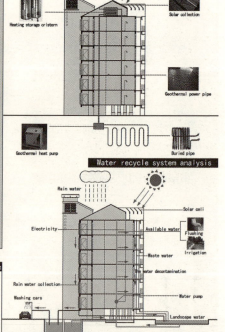

Water recycle system analysis

Water recycle system can retain the rainwater or sink into the ground, improve the water and ecological environment. It also effectively increases the water resources, therefore it is a good measure of saving water. In addition, the wastewater will be filter in sewage pipe of indoor which is buried in the bottom of the pond through the media, and may be directly discharged into the river, effectively reduced the pollution of the river, played a role in environmental protection.

Photovoltaic system

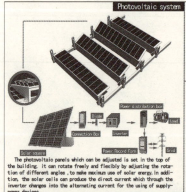

The photovoltaic panels which can be adjusted is set in the top of the building, it can rotate freely and flexibly by adjusting the rotation of different angles, to make maximum use of solar energy. In addition, the solar cells can produce the direct current which through the inverter changes into the alternating current for the using of supply-power devices.

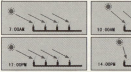

Stairwell ventilation

The chimney of pulling the wind can use air pressure to make indoor air along with a vertical slope to the rise or fall, causes the air to enhance convection. When the indoor temperature is higher than the outdoor temperature. Due to lower density, the hot air of indoor naturally increases along the vertical channel, leaks from the top part through the gap of the doors and windows and all kinds of holes. Due to higher density, the cold air of outdoor infiltrates to add from the low-level it forms "the chimney effect", so achieves the purpose of natural ventilation.

lamp of solar

The lamp of solar makes use of the sunlight as energy, during the day the solar panels charges up the battery, the battery supplies power to the load using at night, without the complex and expensive laying pipeline, the layout of light can be freely adjusted and reliable, saving electricity, no maintenance.

Noiseproof screen

The barrier of noise is a good effect of reducing noisy, the higher structure safety, the anti-strong forces of nature and man-made destruction, the saving investment, the faster speed of construction, the role of significant landscape and so on.

Solar lamp lured insects

The trap lamp of solar energy, which involves that device uses light to trap insect. The lights can frequently work in night of the harmful insect activity, which has some benefits including energy saving, easy to install, easy to manage.

Western-wall planting

The three-dimensional afforestation and shading technology is adopted in the western wall, the vine which twines on a fixed component in the wall, it can absorb a large amount of solar radiation in summer, reduce the temperature of the wall. It also can effectively prevent that the west wall is heated.

The construction of planting proof

Layers: Planting layer; Nutritious soil layer; Isolation filter layer; Drainage layer; Resistance to root puncture waterproof layer; Coil or smear waterproof layer; layer; Structure; Drain hole

SOLAR ENERGY ARCHITECTURE DESIGN — 03

LIVEABLE AND LOW-CARBON RESIDENCE IN WUJIANG

REGISTER NUMBER:1607

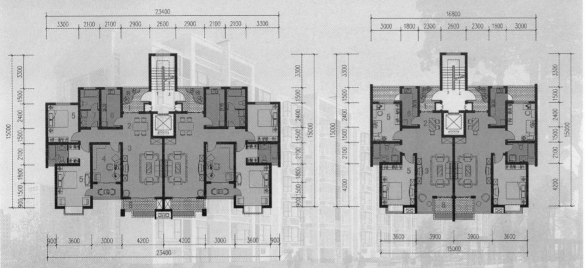

Unit plan 1:100

1. Garden of entrance 2. Dining room 3. Living room 4. Study room 5. Bed room 6. Kitchen 7. Bathroom 8. Balcony

90m² section view

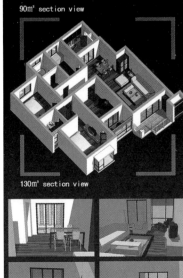

130m² section view

Interior perspective

Solar collection design

The Wall-mounted solar water heater uses the superconducting heat-pipe collector, has some characteristics including the fast spreading, the lower temperature starting, the higher thermal efficiency, the longer life, the anti-cold and so on. In this scenario, let it to combine with the balcony and facade of louvers, to form a unique architectural detail on the texture.

Working principle

Wall hung solar water heater collectors installed outside, storage tank installation indoors, the circulation line connecting cistern and collectors, relying on a complete system temperature sensing sensor forced circulation, satisfying the customer all-weather quantitative or thermostatic water demands. And has the freezing, prevent electricity, lightning protection measures to ensure system and reliable operation, safe use.

Solar house analysis

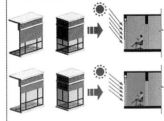

In summer, laying down the blinds, could effectively block the sunlight. In winter, laying up the blinds, can fully receive the sunlight.

Floor heating

The floor heating is that hot water pipes are embedded in the concrete of the ground floor, makes use of the storage thermal capacity and good thermal stability of concrete, increasing the temperature indoor, to achieve the purpose of heating.

Construction of wall

10 thick surface layer
20 thick leveling layer
180 thick wall of earth embryo
180 thick straw board
120 thick autoclaved fly ash brick
20 thick leveling layer
10 thick surface layer

The straw of crop as the main materials is used to produce a new type of wall materials-lightweight straw wall panels, has some advantages, including waterproof, fireproof, shockproof, anti-crack, anti-aging, acid and alkali resistant, anti-shock, no radiation and so on. it takes full advantage of the straw, reeds, big poplar seedlings, corn stalks, rice husks, sawdust and so on, can turn rubish into treasure.

Nature ventilation analysis

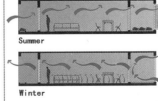

Summer

Winter

In summer, opening all the doors and windows and grille shutters of home garden, it can create a good drafts. In winter, closing the louvers and doors, windows, can use the inducing natural ventilation technology of pulling wind

Lured ventilation

Heat insulation window

LE glass with the metal coated, can reflect the indoor thermal radiation back to the room in winter, ensure that the heating of the indoors is not lost toward the outdoors. it can prevent the thermal radiation from the sun of outdoor into the room in summer.

SOLAR ENERGY ARCHITECTURE DESIGN   04

优秀奖
Honorable Mention Prize

项目名称：土地"改革"
Land Reform

作　者：吕扬伟、何海峰、宋　廉、
王修水

参赛单位：浙江树人大学城建学院

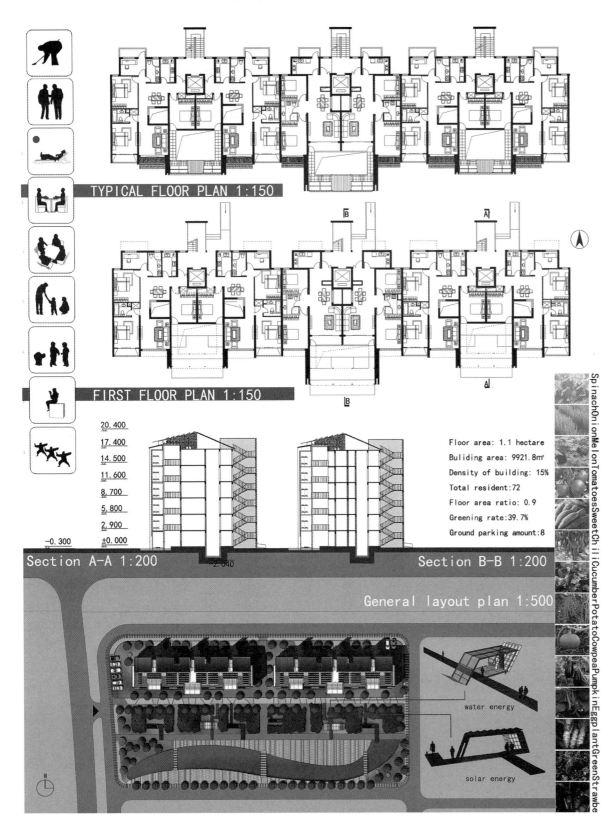

Land Reform — Growing In The Sunshine

1617

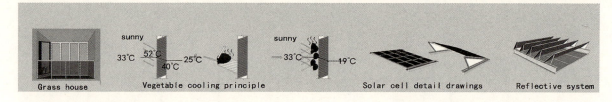

MAIN STRATEGY

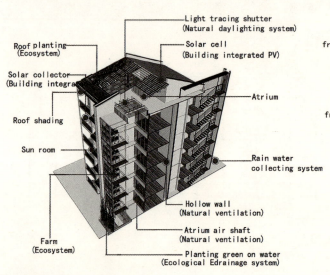

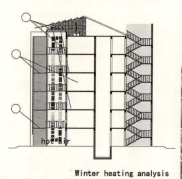

TECHNICAL ANALYSIS

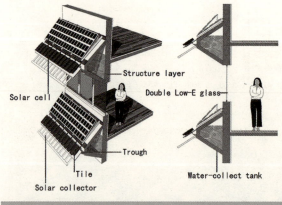

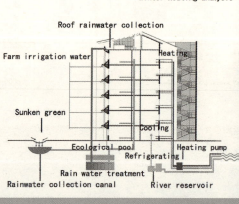

SOLAR HOUSE ANALYSIS

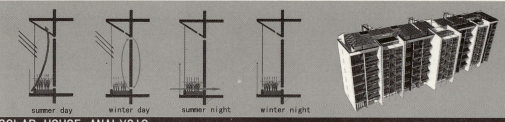

Land Reform — Growing In The Sunshine

Your neighbor, your friends

Our "land"

The plan of our farm

The section of our farm

Portable farm

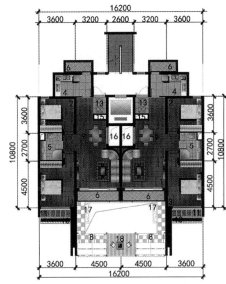

Dwelling Size A

Dwelling Size B

1. Living room
2. Bedroom
3. Chief bedroom
4. Kitchen
5. Bathroom
6. Balcony
7. Grass house
8. Farm
9. Dining room
10. Solar collector
11. Solar cell
12. Study room
13. Courtyard
14. Porch
15. WSHP pipe
16. Inside patio
17. AC
18. Sharing space

SpinachOnionMelonTomatoesSweetChilliCucumberPotatoCowpeaPumpkinEggplantGreenStra...

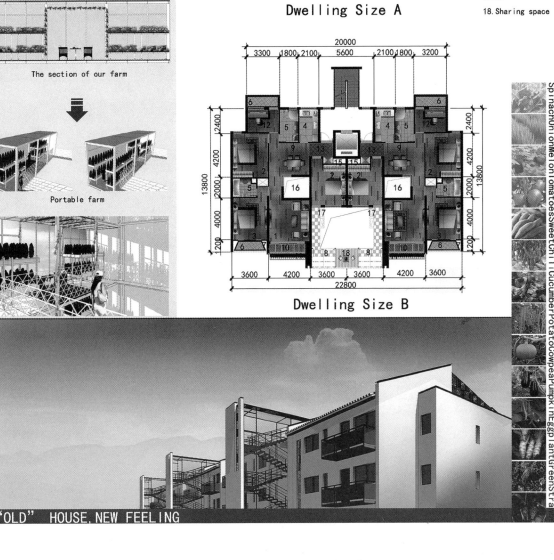

"OLD" HOUSE, NEW FEELING

优秀奖
Honorable Mention Prize

项目名称：地球24小时
　　　　　Earth-24H
作　者：张　日、黄杉杉、苏　平、
　　　　陈　波、叶尚晶、俞立卫、
　　　　黄裕镯、王修水
参赛单位：浙江树人大学城建学院

Earth-24H WUJIANG LOW CARBON LIVABLE RESIDENCE [1]

General analysis

Tongli Town is located in the northeast of Wujiang City, Jiangsu Province, is a long history and the typical style of the ancient town of rivers and lakes, we try to use the modern design techniques to integrate the new buildings into this historic and become a part of it.

Concept analysis

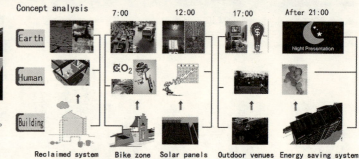

7:00　12:00　17:00　After 21:00

Reclaimed system　Bike zone　Solar panels　Outdoor venues　Energy saving system

设计说明：

地球是人类赖以生存的家园，为人类灿烂文明提供物质基础，而文明的背后也正是人类自身活动侵蚀了时间所见证的历史文明留下的烙印，日出而作，日落而息，在这24h里人类在生产物质的同时也在消耗物质，在稍纵即逝的24h里，人类有必要思考我们的行为模式——低碳生活。传承文明，呵护地球，从这简单的24h开始做起……

Design report:

Earth is the home to human life, to provide the material basis of human splendid civilization, but behind the civilization. The human activities destroy their own civilization which has created by themselves, start with sunrise and stop at sunset, in this 24 hours, the human are both producing and depleting the materials, in this fleeting 24hours, we need to think about our behavior—low carbon life. Heritage of civilization, take care of the earth, let's start from this simple 24 hours...

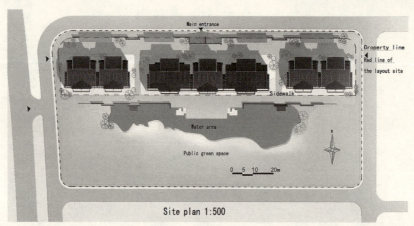

Site plan 1:500

Economic index

No.	Name	Unit	Quantity
1	total land area	Ha	0.011
2	gross building area	m²	10096.2
3	building density	%	37.7%
4	total rooms	rooms	84
5	floor area ratio		2.6
6	greening rate	%	31.4%
7	parking rate	%	14.3%

Southeast wind　→Noise　Sound-insulated wall
Northwest wind
　　　　　　　　Green space
　　　　　　　　Waterscape
　　　　　　　　Observatory
　　　　　　　　Axes
● 90m²　→Road
● 130m²　↔ Pedestrian road

WUJIANG LOW CARBON LIVABLE RESIDENCE [2]

[steel structure]

The first is can be dry construction, water conservation; less construction area, less noise, less dust
The second is less soil used in foundation construction because of self-weight reducing, less destruction of land which is a valuable resource.
The third is a significant reduction in using the concrete and tiles, reduce to dig stones from mountains around cities, conducive to environmental protection.
The fourth is when the building is expired, less solid waste from removing structure, the price of recycling the waste steel is high.
The fifth is reducing the size of structure, raise the room using rate.

1. Reduce the banding workload, shorten the construction period
2. Less waste, save materials
3. Construct several floors at the same time
4. Reinforced arrangement and improve the construction quality evenly

1. Green, Environmental Protection (Use the sand tailings and other wastes as raw materials)
2. Rapid construction
3. Wall formation, without plastering saving material and human
4. Light weight, high strength

Solar collector panels 1: to provide daily hot water
Sunshine at the north window: to improve the quality of the life in it
Photovoltaic panels: to provide daily supply
Solar collector panels 2: to provide space heating and wind power pull
Fresh air device: to provide the fresh air
External walls: reducing the noise urban roads
Parking area
Bike zone

Solar collector panels 3: to heat the fresh air and Promote the circulation of fresh air
Insulated joints: separate thermal conductivity of the balcony
Landscape Wall: the looming garden style of the building
Vents: Ventilation and Moisture-proof
Observatory: inherit the features

Beam-column connections use the through-type rigid connection

Interior Wall   Outer Wall   Outer Wall

Concept of low-carbon buildings

[Aerial view]

1. master bedroom 3. kitchen 5. dining room 7. living room
2. bathroom 4. balcony 6. equipment room 8. second room

[Floor plan]
90B UNIT CELL PLAN 1:100
TYPICAL FLOOR PLAN 1:300
SIXTH FLOOR PLAN 1:300
TYPICAL FLOOR PLAN 1:300
TOP FLOOR PLAN 1:300
[Floor plan]

[North elevation]

Earth-24H WUJIANG LOW CARBON LIVABLE RESIDENCE [3]

Earth-24H WUJIANG LOW CARBON LIVABLE RESIDENCE [4]

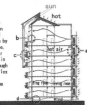
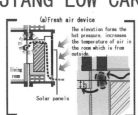
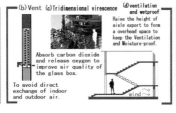
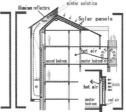

[Winter]

In the winter, the air is heated by the hot plate in the fresh air device, and then enters into the room by following the hot pressure, provide the warm fresh air and keep the heath index in the room, and the sun through aluminum reflex plate reflex the light onto north windows, improve north room living environment.

(a) Fresh air device — The elevation forms the hot pressure, increases the temperature of air in the room which is from outside.

(b) Vent (c) Tridimensional virescence — Absorb carbon dioxide and release oxygen to improve air quality of the glass box. To avoid direct exchange of indoor and outdoor air.

(d) ventilation and wetproof — Raise the height of aisle export to form a overhead space to keep the Ventilation and Moisture-proof.

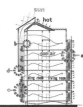
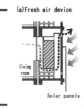
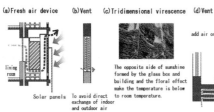
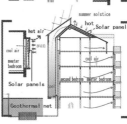

[Summer]

In the summer, The geothermy provides the cool wind to glass box under the hot pressure, the air goes up after being heated by the hot plate in the fresh air device, causes to form the negative pressure in the room, the cool healthy air in the glass box will enter in and keep the healthy index in the room.

(a) Fresh air device
(b) Vent (c) Tridimensional virescence — The opposite side of sunshine formed by the glass box and building and the floral effect make the temperature is below to room temperature.
(d) Vent — add air or leak air

[Constitute]

[Balcony insulation]

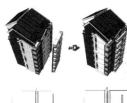

The traditional approach / Our approach

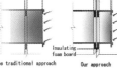

Disengage the balcony and main building and set a seam, to form two independent systems in order to cut off the balcony on the building's heat conduction, make the balcony becomes the main architecture sunshade.

[Rain water collection]

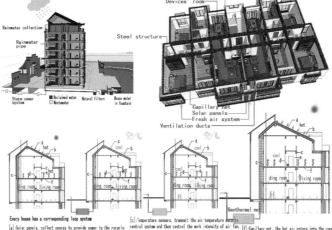

Every house has a corresponding loop system

[a]: Solar panels, collect energy to provide power to the recycle.
[b]: Photovoltaic panels, collect the solar energy to supply energy to every house.
[c]: Temperature sensors, transmit the air temperature data to central system and then control the work intensity of air fan.
[d]: Air fan, to provide power to the air cycle.
[e]: Central system equipment room
[f]: Capillary net, the hot air enters into the room uniformly through capillary net which is embedded in floor, and prevents the smell into the room.

优秀奖
Honorable Mention Prize

项目名称：水意阳情
　　　　　Solar & Sustainable
　　　　　Residence Design

作　者：徐燊、廖维、殷实
参赛单位：华中科技大学建筑与
　　　　　城市规划学院

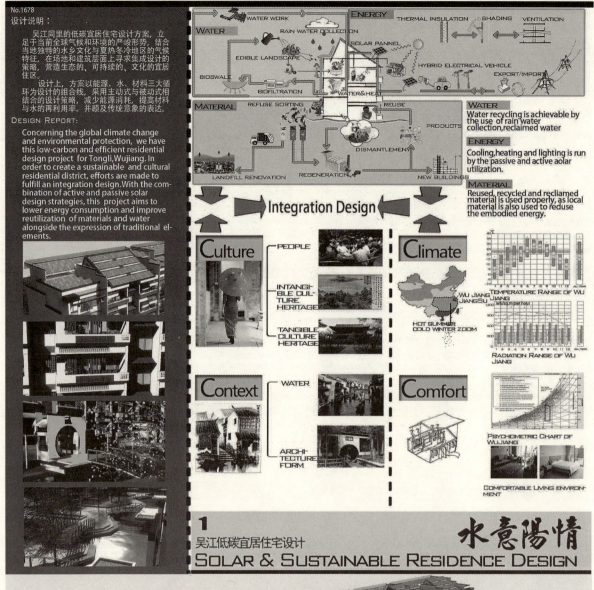

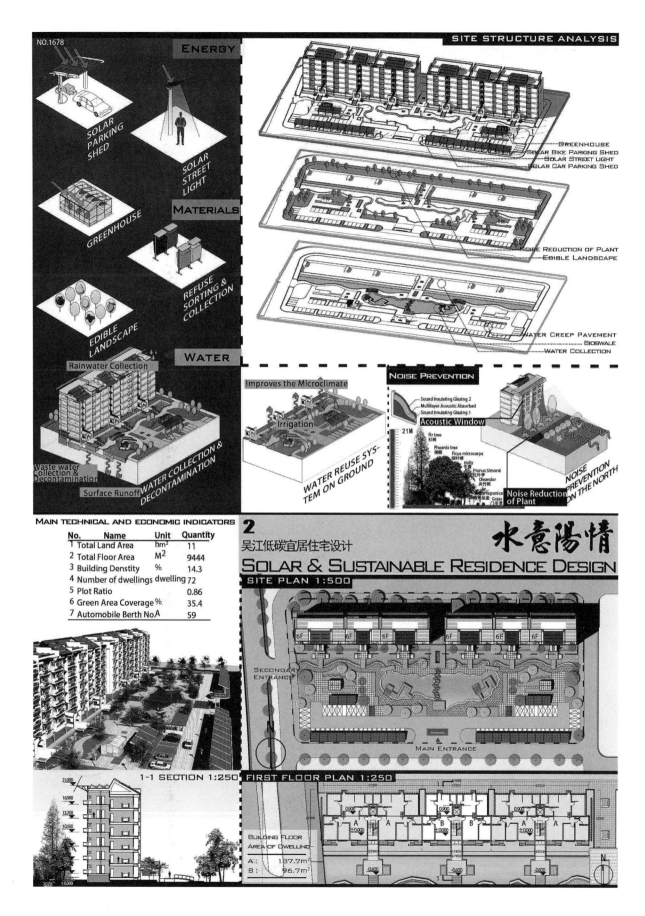

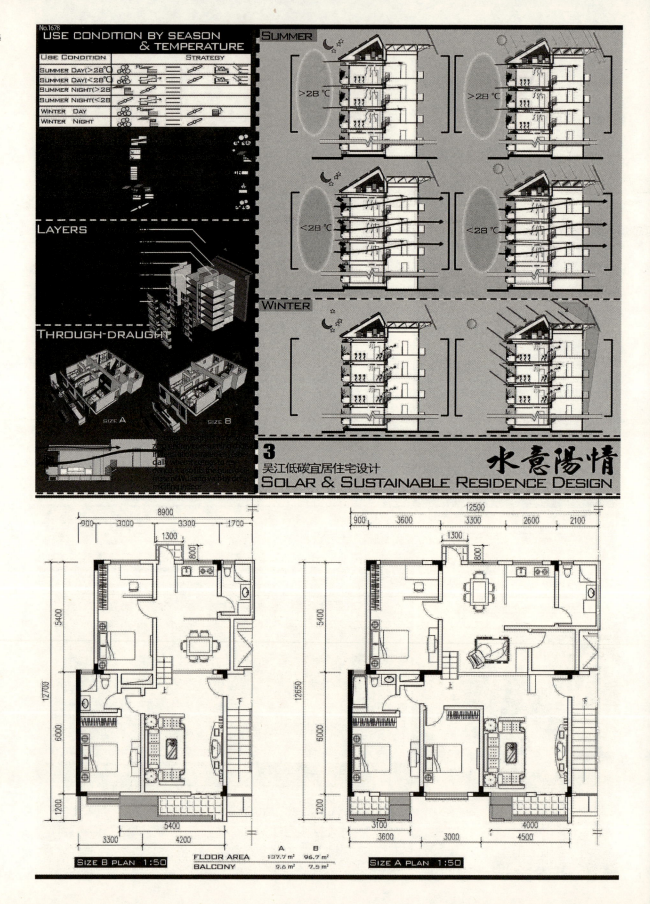

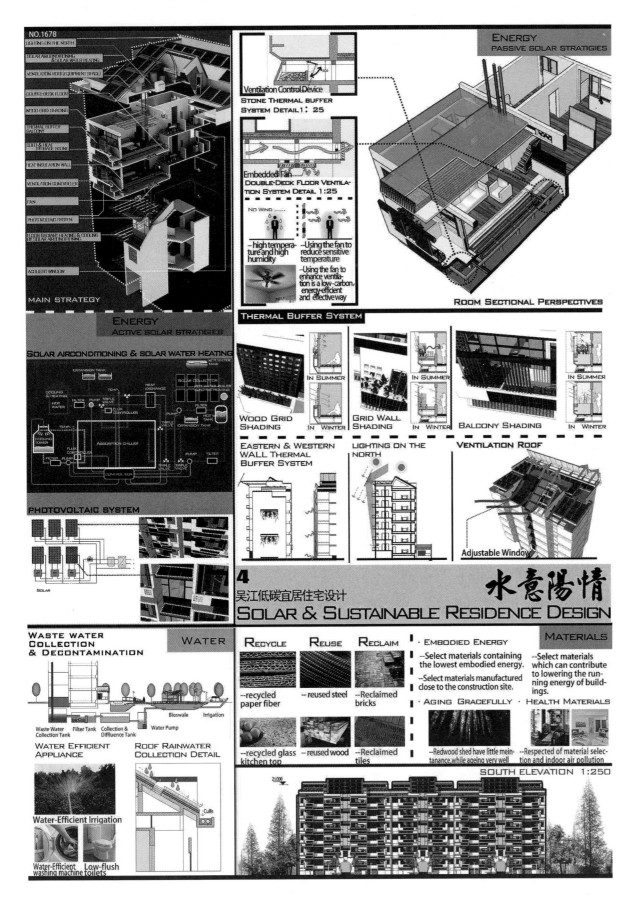

优秀奖
Honorable Mention Prize

项目名称：山居
　　　　　Shan Ju
作　　者：李　静、何明星
参赛单位：南阳师范学院土木建筑工程学院

1816　山居 No.1
Low carbon green imagination of solar architecture

climate character of wujiang

Wujiang city located in the Yangtze river downstream taihu lake basin in China building thermotechnical design category of hot summer and cold winter zone, the typical monsoon climate. The temperature difference is small, rain, plump and four distinct seasons. Winter, the summer heat temperature little rain more rain, air humidity is big, annual average reached 79%. Annual average temperature is 15.7 degrees, the average annual rainfall for 1019mm, summer, winter prevalence southeast wind from Mongolia, Siberia inland blowing northwest wind.

current dwelling energy situation

With the rapid development of economy, the energy also unceasingly reduction. But the current energy utilization efficiency is still low, the rational utilization of energy and exploitation of new energy become particularly important.

BUILDING VS POWER STATION

Higher consumption lighting "waste" growth each year equals to shenshen nuclear power station

ENERGY STRATEGY — INCORPORATION

Architecture

TRIDIMENSIONAL VIRESCENCE

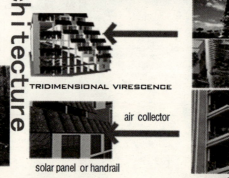

air collector

solar panel or handrail

low-e glass

window

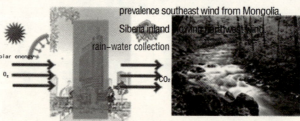

rain-water collection

solar energy
O_2
CO_2

BUILDING VS PLANT

Air conditioning daily consume large amounts of energy, and cannot achieve ideal effect, energy efficiency is quite low.

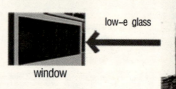

fresh air
4.3 billion kW/h electricity
6.08 million hot water

BUILDING VS MACHINERY

设计说明：

吴江市低碳宜居住宅规划设计，位于同里文化古镇周边。本设计考虑文化底蕴、基地自然环境与气候条件，从总体布局到建筑设计注重以太阳能和雨水收集为主，进行综合利用。人们面对资源匮乏与土地的缺乏问题要从多方面考虑节约能源以开能源，同时采用垂直空间的方法解决土地缺乏问题。将农田与住宅相结合与同里的生活文化符，同时又因调研发现建筑用地原是农田。本计采用阶梯状的组合方式，利用剩余的屋顶空间使每户都有庭院，与此同时还将同里古建筑的特色运用于本建筑，营造舒适的生活空间。

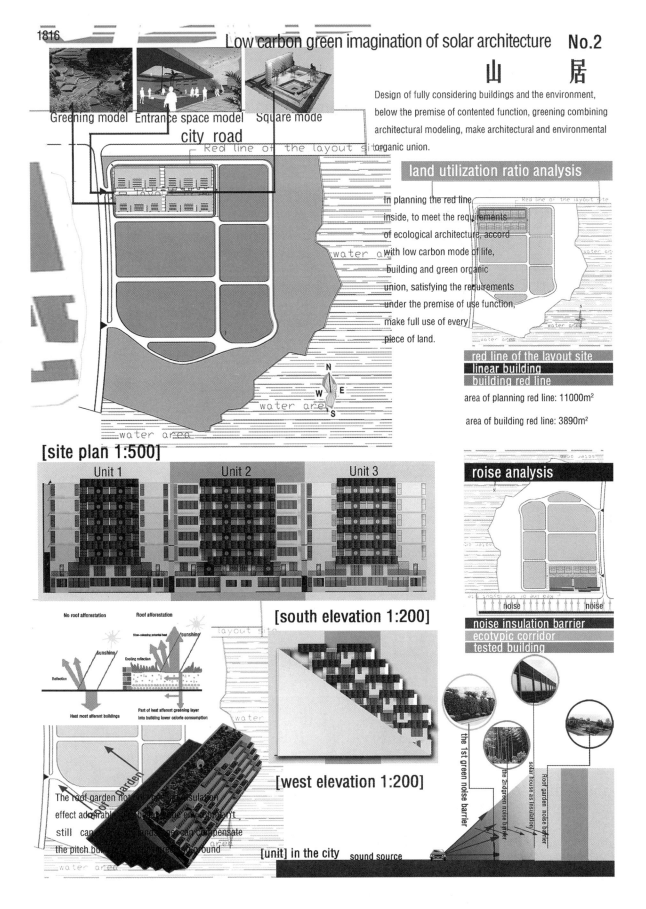

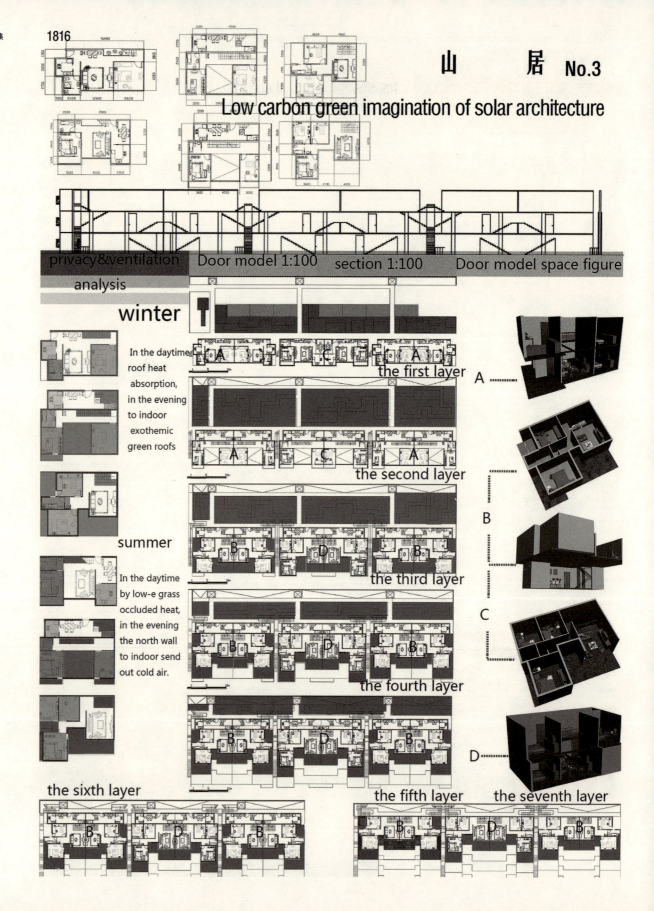

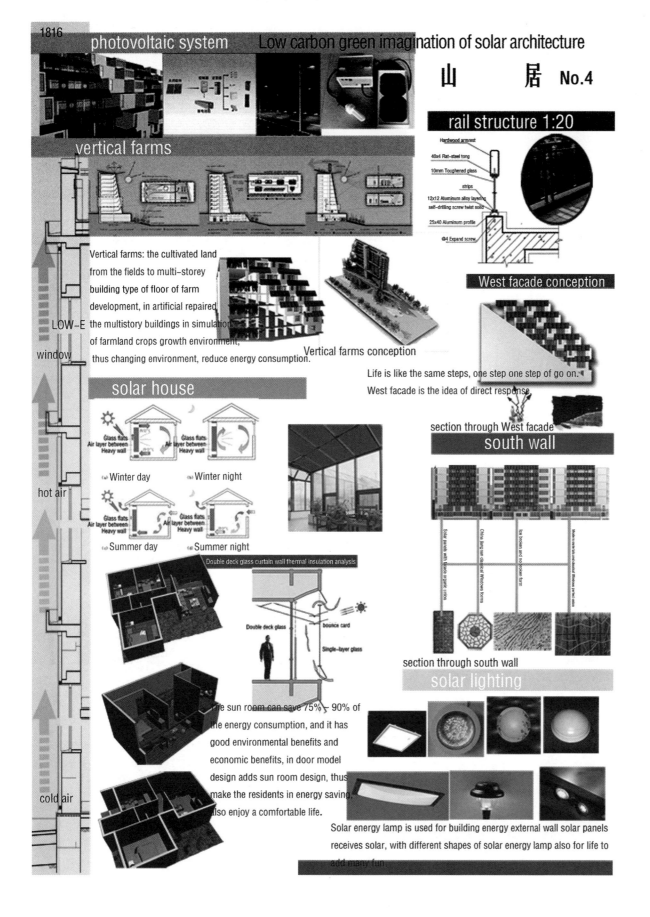

优秀奖
Honorable Mention Prize

项目名称：阳光家园·江南情怀
Sunshine · Water · Home

作　者：胡南江、蔡　宁、黄祖坚、
　　　　　钱乔峰、袁小雨、张　智、
　　　　　龚　哲、蒋钧海、张宇峰、
　　　　　蔡　健

参赛单位：华南理工大学建筑学院

注册号：1999

Local Climate and Geographical Characteristics
black and whigt, sloping

Local Architectural Feature
qiuet, harmonious, joy, satisefy

Local Human Cultrue
water, boat, reflection, clear, flowing

设计说明：吴江市低碳宜居住宅设计，着眼于当地资源、环境、气候、民俗等的分析，通过多种技术手段以降低建筑能耗，节约环境资源，利用清洁能源，达到建设低碳建筑的目的。同时，由于吴江独特的地理位置条件，设计中重点用了苏州地区古民居的建筑符号，将传统元素融入现代建筑，充分尊重了古镇的水乡风貌及当地的民居特色。

Design Report: The design of low-carbon house in Wujiang city is founded upon the analysis of actual situation (resource, environment, climate, tradition custom etc.), focusing on making use of solar energy and wind energy resource and utilizing many technique to decrease the consume of the architecture, save the environmental resource, make full use of the clean energy to reach the goal of low-carbon design. At the same time, the design implants the traditional architecture elements, and merges it in the low-carbon house, respecting the style and feature of the traditional old watery town.

WIND SPEED　TEMPERATURE　SUN RADIANT　HUMIDITY

- city road
- influence of the noise
- landscape point
- obit of the sun
- joint of the trasportation

⇒ summer wind insulation
⇒ winter wind insulation

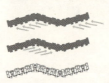

Architecture
Main Body Design
The architecture main body is tilted 15° and in a folded line to gain the best insolation condition and utilise the solar energy. In this way, the architecture not only gets a reduction of shape coefficient and waste of energy, but also enlarges the scenery surface. In the same time, it can fully utilise the estern and western sunshine and save it in the thermal storage clapboard and floor.

Economic and Technical Indicators

Name	Area	Amount
The total land area	m²	2134.97 m²
The total construction area	m²	10644.27+891.83 =11536.1 m²	(283.74+279.87+203*5 =1578.61 m² each floor)
Overall land area	m²	3890
Overall building area	m²	11000
building density	%	19.41	2134.97/11000
Total households		82
Floor area ratio		1.049	11536.1/11000
Green rate	%	36.7
Car parking rate	%	75

Name	Area (m²)
House area of A	130
Entrance Garden	12.1
Living Room	26.1
The Sun Room	6.6
Dining Room	12.1
Kitchen	7.4
Master Bedroom	24.2
Second Bedroom	16.3
Second Bedroom 2	13.2
Toilet	4.8
Corridor	6.5
The Total Area	129.3

Name	Area (m²)
House area of B	90
Living Room	24.8
The Sun Room	6.8
Dining Room	8.7
Kitchen	5.7
Master Bedroom	21.8
Second Bedroom	19.6
Toilet	3.8
The Total Area	90.2

阳光家园·江南情怀　吴江市低碳宜居住宅设计

SUNSHINE HOUSE DESIGN

Concepetual Design　vol.1

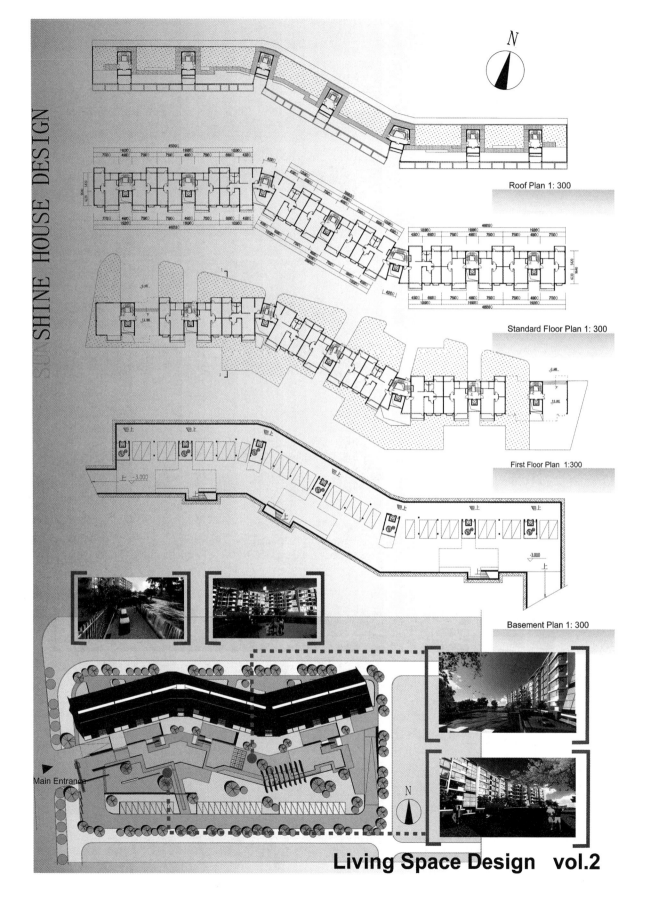

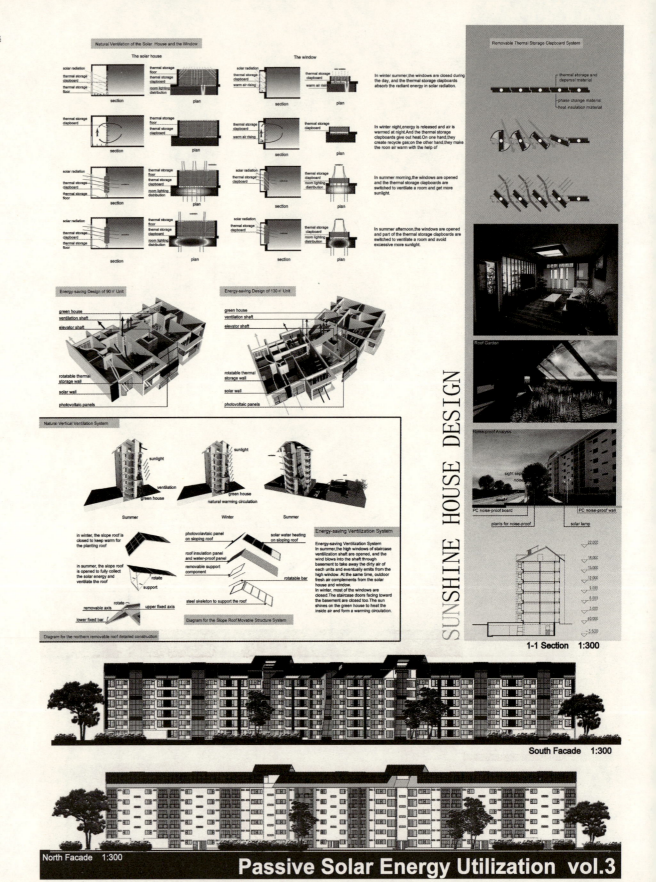

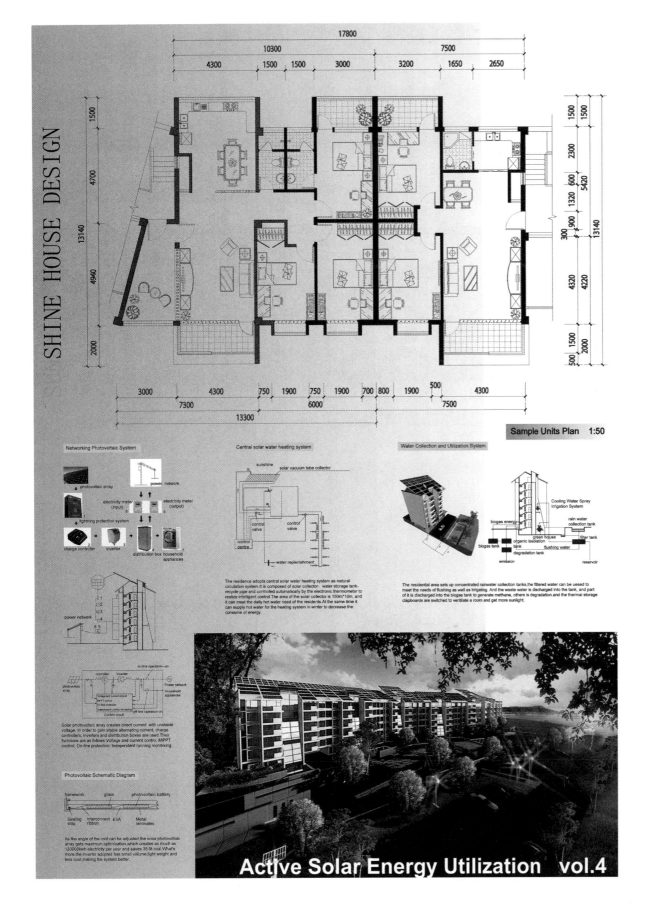

优秀奖
Honorable Mention Prize

项目名称：沐姑苏
　　　　　Livable and Low-carbon
　　　　　Residential Building
　　　　　Design in Wujiang
作　　者：孙倩倩、任娜娜
参赛单位：山东建筑大学建筑城规学院

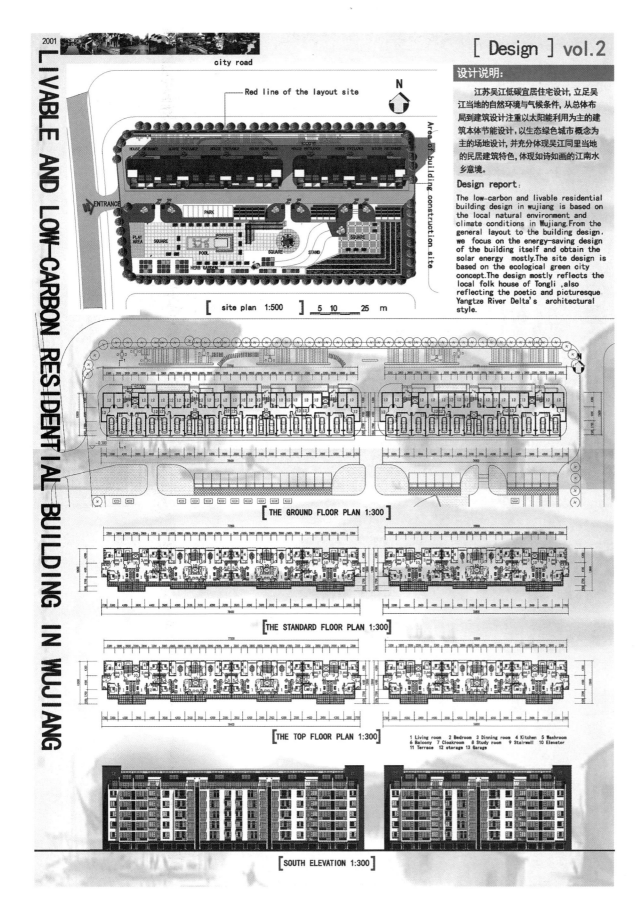

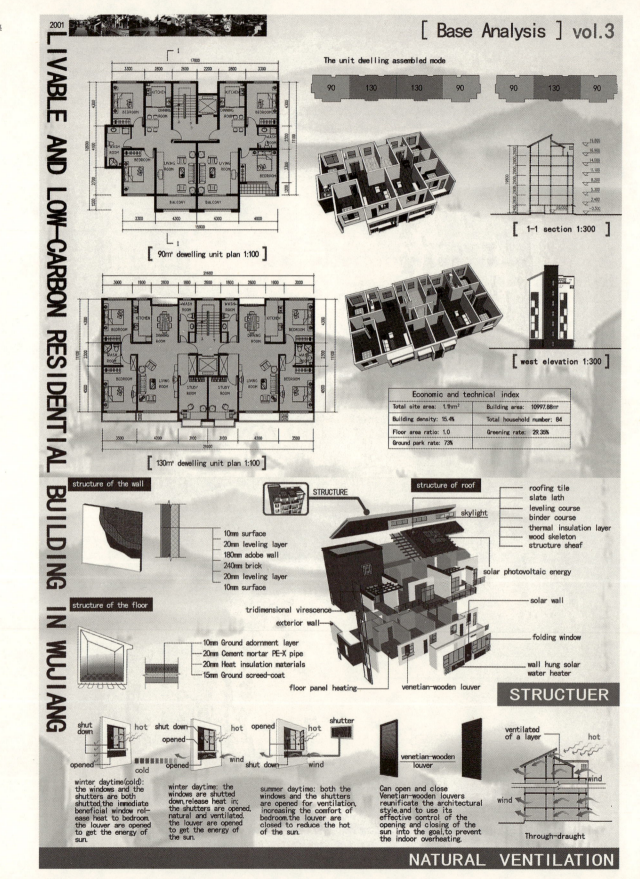

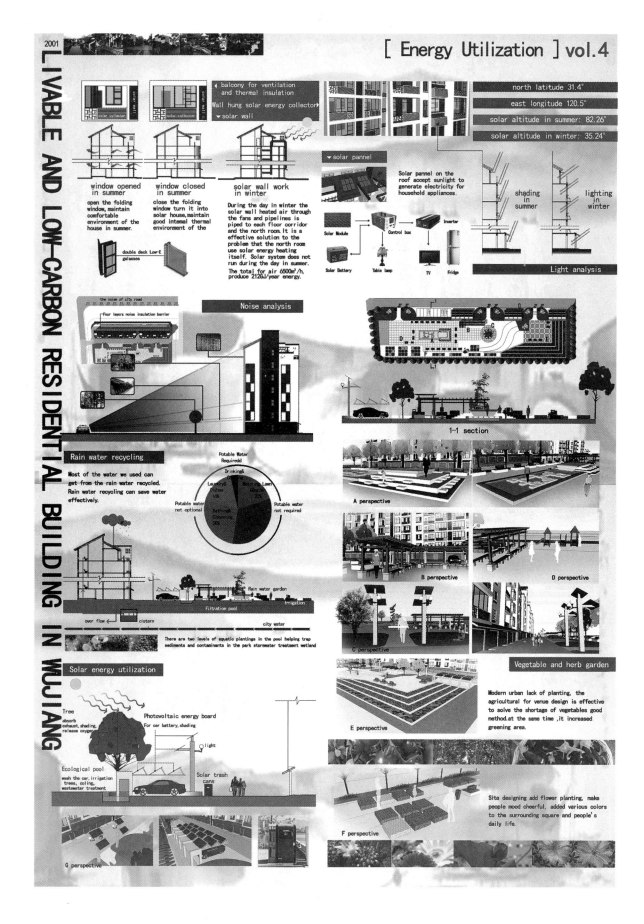

优秀奖
Honorable Mention Prize

项目名称：风光无限
　　　　　Limitless Wind Limitless Solar
作　者：高力强、欧阳仕淮、宋宏浩、
　　　　王　帅、张祎娇、徐禛龙、
　　　　姚　瑶、姚辰明
参赛单位：石家庄铁道大学建筑与艺术学院

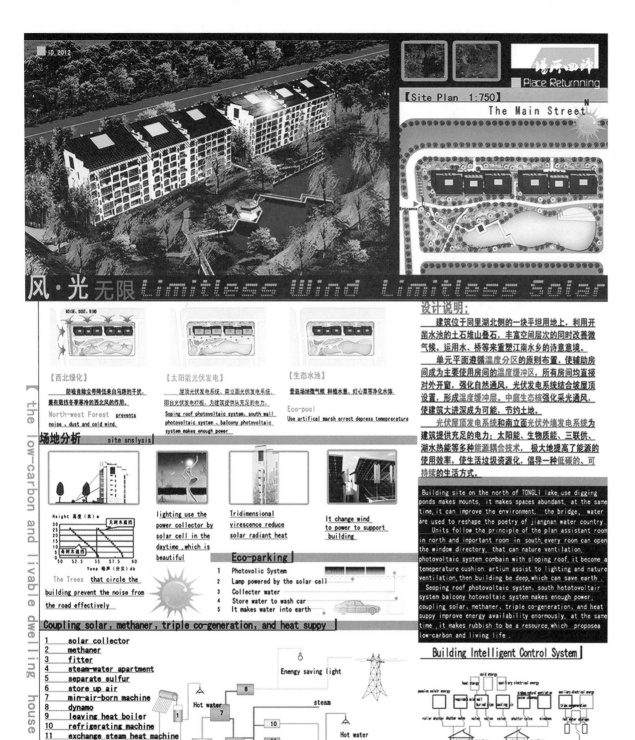

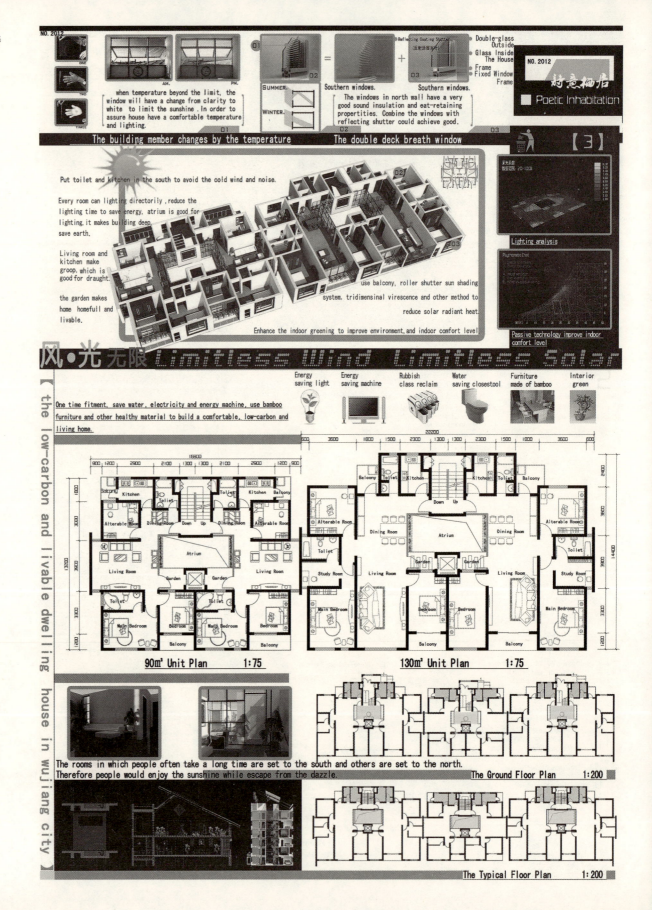

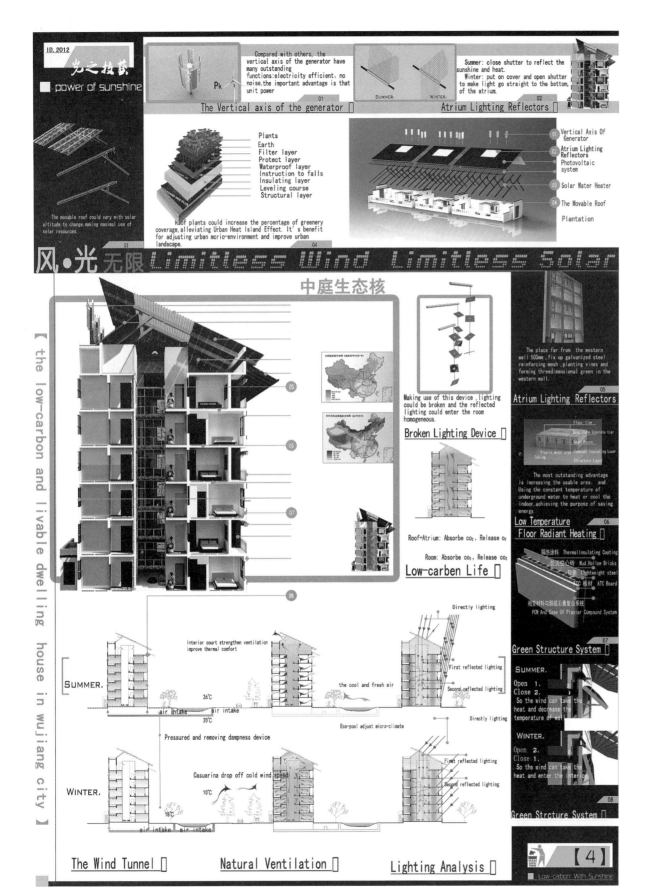

优秀奖
Honorable Mention Prize

项目名称：归园恬居
　　　　　Low-carbon Residence Design
作　　者：曲文昕、张子涛、雷　阳、
　　　　　刘　夔　曲文晓
参赛单位：山东建筑大学建筑城规学院、
　　　　　青岛理工大学建筑学院

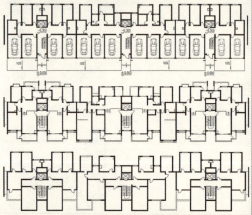

Design specification

Originating in the theme of "sunshine and low-carbon life", we design a resident environment which is comfortable, economical and environmental. We inspired from Suzhou traditional residence and Tongli Town that is implicative and spontaneous, to create a rhythm and readability for the community. Meanwhile, we guarantee comfort and percentage of usable dwelling space.

ventilation and dehumidification system which include solar chimney and ground source heat pump build a cool and dry indoor environment by providing fresh air. Heat insulation wall and adjustable sun shading board block the blazing sun in summer and cold breeze in winter. Solar-ground source heat pump, chilled ceiling and radiant floor heating system constitute heating and cooling system which is highly efficient, clean and environmental-friendly.

"A trapped bird longs for its forest, a pool fish misses its puddle." As we trapped in the city forest, a "dwelling of heart" may bring us the feel that "the nature backs and my heart to follow".

设计以"阳光与低碳生活"为主题，来营造"舒适、经济、环保"的居住环境。设计从苏州民居和同里古镇的意蕴出发，含蓄而自然，同时又保证了户型空间的实用舒适和高利用率，完整而细致。一套结合太阳能热烟囱和地冷管技术的通风除湿系统，在夏季带给每户清凉干燥的新风；隔温墙和可调节遮阳板既可以阻挡夏季赤热的阳光，也可以在冬季不遮挡阳光的同时阻挡寒风；太阳能热泵与冷辐射顶棚和地暖结合成为制冷和供暖系统，高效、卫生、环保。

"羁鸟恋旧林，池鱼思故渊"，在如今紧张的城市环境中，这处"恬居"或许能带给人"复得返自然"的清新感受。

NO.	Name	Unit	Quantity	NO.	Name	Unit	Quantity
1	total land area	hm²	1.1	5	volume fraction		0.833
2	total building area	m²	9168	6	green area coverage	%	67.75
3	building density	%	12.55	7	parking proportion	%	53.33
4	total amount		60				

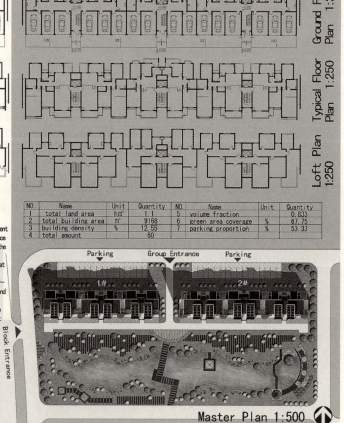

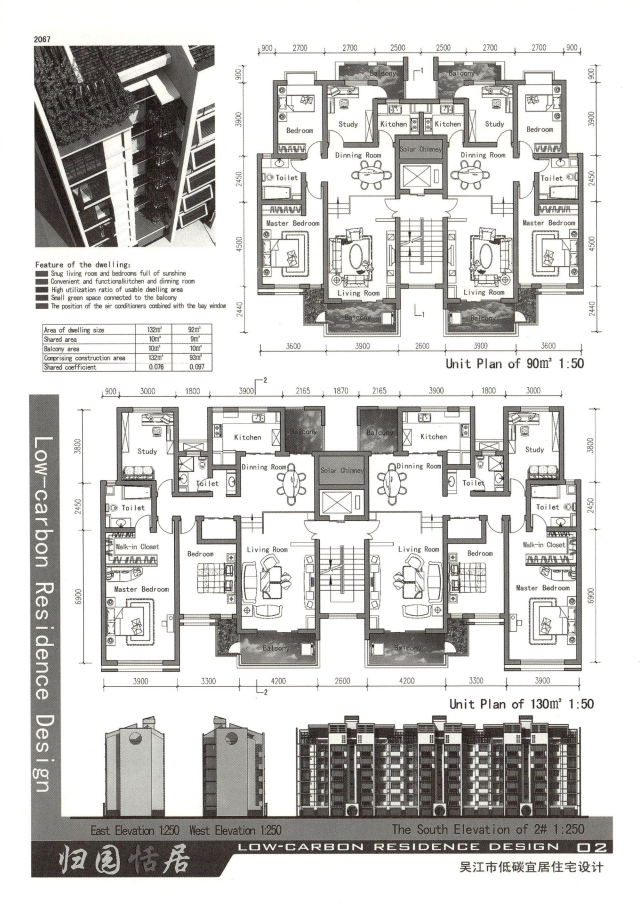

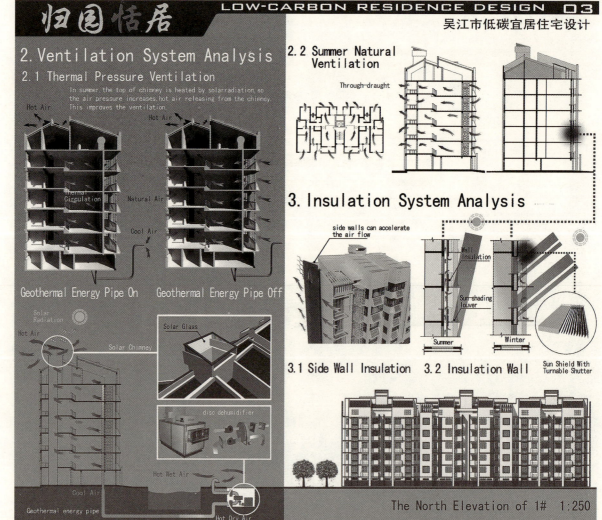

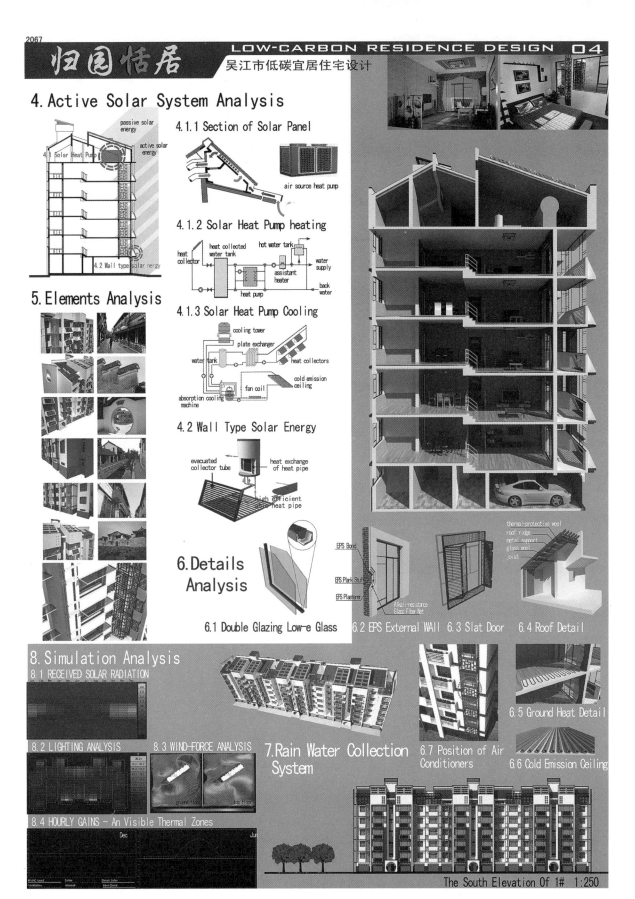

2011 台达杯国际太阳能建筑设计竞赛获奖作品集

优秀奖
Honorable Mention Prize

项目名称：光·享住宅
　　　　　Sunlight Enjoying Residence
作　　者：徐跃家、梁辰、胡英
参赛单位：大连理工大学建筑与艺术学院

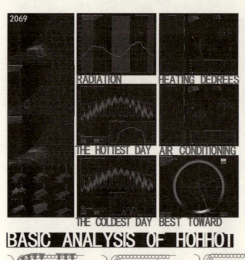

SITE PLAN 1:500

BASIC ANALYSIS OF HOHHOT

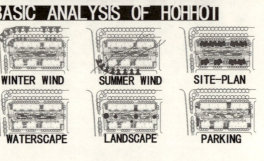

WINTER WIND　SUMMER WIND　SITE-PLAN　VENTILATION
WATERSCAPE　LANDSCAPE　PARKING

TECHNICAL & ECONOMIC INDEXES	
FLOOR AREA	1.1 hm²
BUILDING AREA	23700 m²
BUILDING DENSITY	20.6%
FLOOR AREA RATIO	2.15
RESIDEBTIAL HOUSEHOLDS	280
VIRESCENCE RATE	39.0%
PARKING RATE	63.6%

PROGRAMMING ANALYSIS

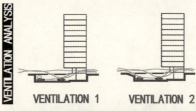
VENTILATION ANALYSIS
VENTILATION 1　VENTILATION 2

SUNSHINE ANALYSIS
SUMMER SUNLIGHT　WINTER SUNLIGHT

UNDERGRAND PLAN 1:1000

SUNSHINE ENJOYING SOLAR RESIDENCE
光·享 住宅

Residential is located in Hohhot.

Programming level: The main rooms of the building face south. The dot-type building sit before the platbuilding as well as the short building sitting before the tall building. The car can get into the building both on the ground and underground. Cars and people are shunted. There are holes for the underground floor daylighting which is good for community ventilation and landscape.

Building level: One stair services three households. The stairwells is designed as a solar chimney. The reflecting board is set for the building. People can get to the roof afforestation which is good for the ventilation. Integrated roof solar collectors, new wind wells, rain collecting system are set too.

Housestyle level: The layout is compact. The space loop is designed to optimize the ventilation and daylight of the house. There are double ventilation floor, 2-layer thermal storage, low-E glass, solar house, shutter etc.

住宅选址呼和浩特市。

规划层面：建筑南北向，前点后板，前低后高布局。场地设地面、地下行车双层入户。

住栋层面：一梯三户。设计太阳能烟囱式楼梯间，北向阳光反射板，可上人通风。

套型层面：布局紧凑，设计空间回路增加户内采光通风。集成双层楼板，双层蓄热墙，双层玻璃窗，阳

WIND ANALYSIS

LANDSCAPE PROGRAMMING ANALYSIS

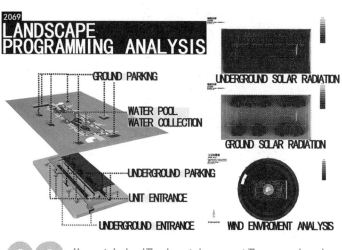

02 UNIT ANALYSIS

Housestyle level: The layout is compact. The space loop is designed to optimize the ventilation and daylight of the house. There are double ventilation floor, 2-layer thermal storage, low-E glass, solar house, shutter etc.

MORNING SUNSHINE / EVENING SUNSHINE / NOON SUNSHINE

SPACE LOOP / A/VENTILATION / B/VENTILATION / C/VENTILATION

FIRST FLOOR UNIT PLAN 1:100

UNIT PLAN 1:50

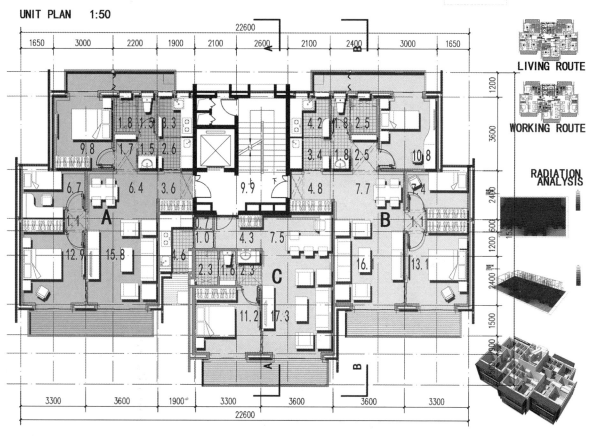

LIVING ROUTE

WORKING ROUTE

RADIATION ANALYSIS

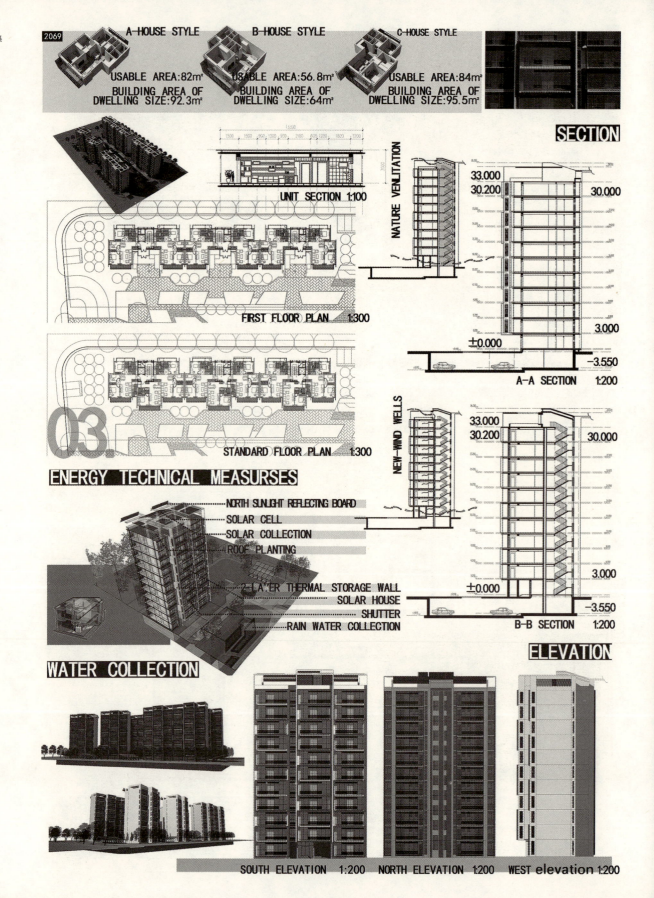

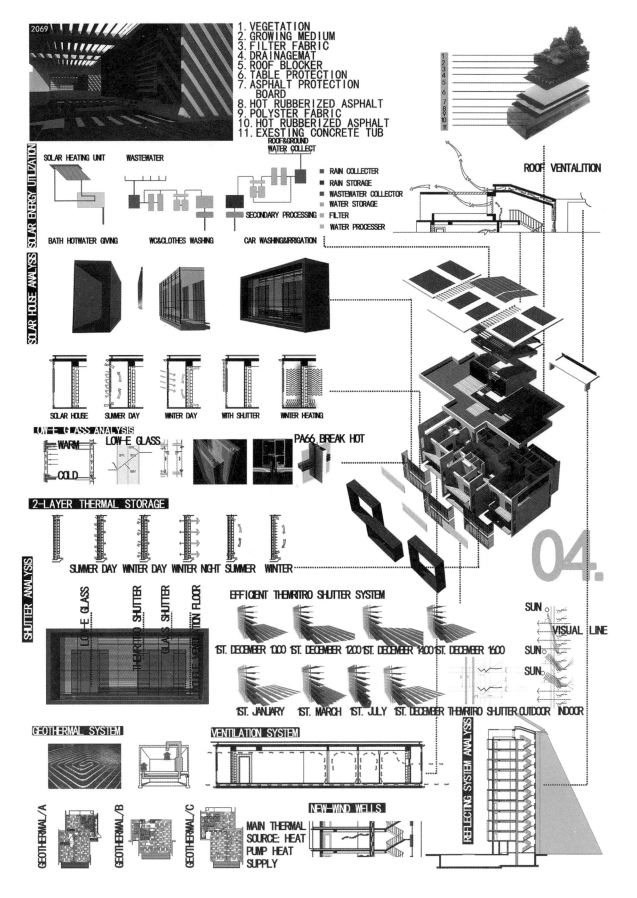

优秀奖
Honorable Mention Prize

项目名称：二分宅
Brother House

作　　者：潘　昊、张愚峰、赵广颖、
　　　　　斯振彬、严　露、叶　枫

参赛单位：武汉科技大学

BROTHER HOUSE 二分宅
CONCEPTUAL DESIGN 1

设计说明

该方案造型上对传统江南民居元素进行解析与重构，传统中具有现代感。住宅户型一分为二，形成前后两进空间，通过连廊进行连接，造成私密空间与公共空间的自然分离。

户与户之间形成由连廊围合而成的天然采光井。在庭院空间的顶部设有中庭反光板，使庭院内部具有良好的采光。建筑局部架空，形成公共活动空间，为住户提供休闲场所。

INSTRUCTION

The program analyses and reconstructs the traditional elements of the dwellings in Southern Yangtze area, in which is modern style with tradition. Every unit is made of two parts, which connected by the corridor and ensure the separation of private space and public space.

There are courtyard spaces formed by the corridors between the families, which creat the natural airshafts. Atrium reflectors are installed on the top of the atrium space, which keep the atrium lighting well.

perspective

SMILING Face...

Location: wujiang is in jiangsu province

What is Low-carbon and Livable residential? It all depends to the people living in the house.

Space Layout: paralle space

Materials: the wall is decorated with wood
the roof is covered by grey clay tile

SOLAR ENERGY ARCHITECTURE DESIGN

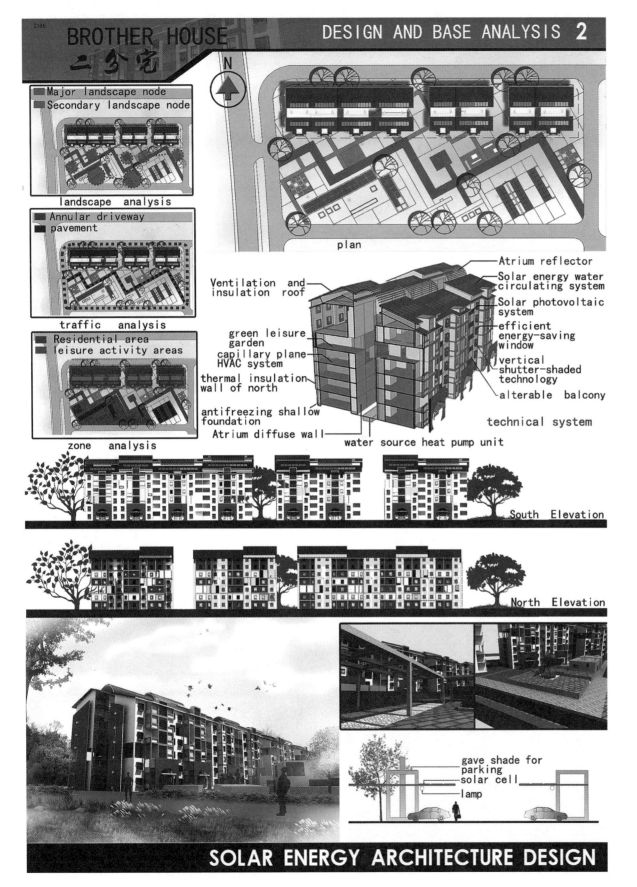

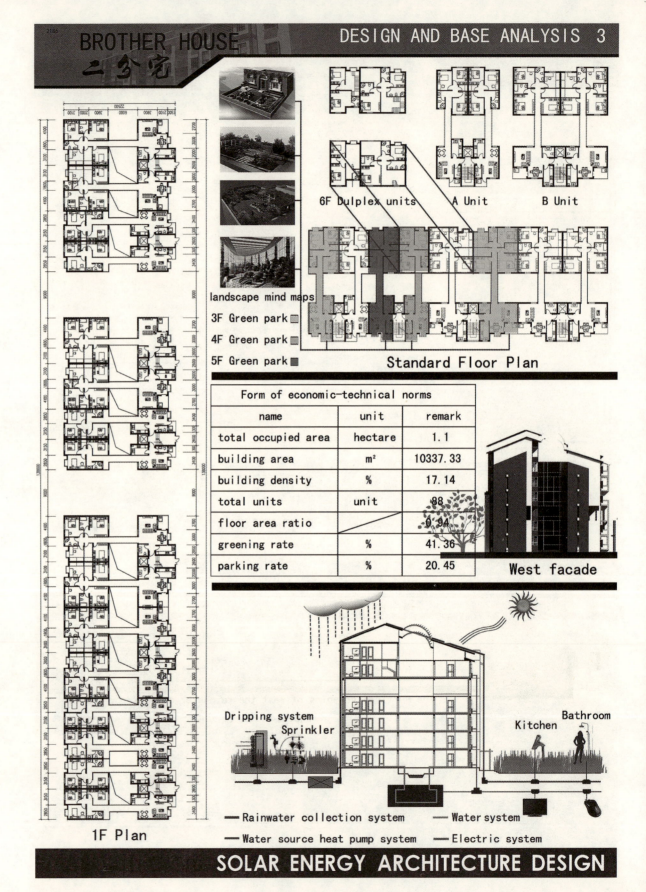

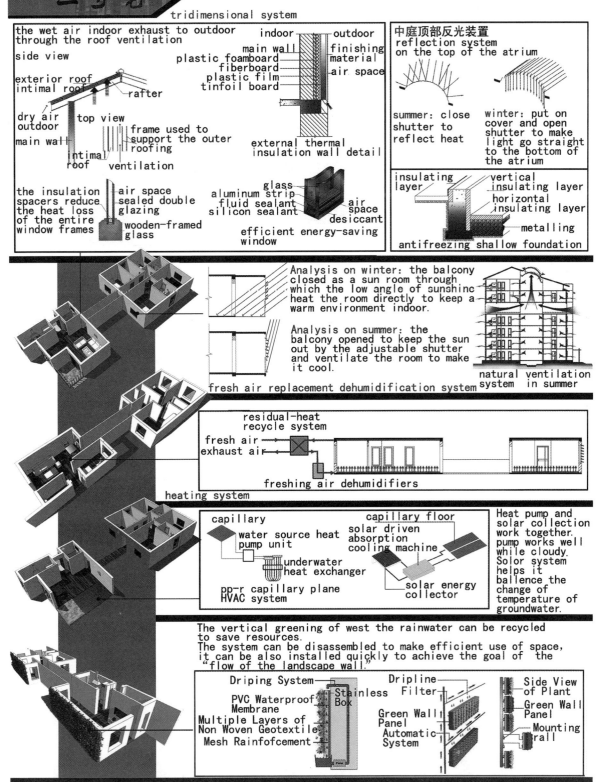

ANALYSIS

优秀奖
Honorable Mention Prize

项目名称：阳光·风
　　　　　Solar·Wind
作　者：范润恬、吴鹏飞、陈未
参赛单位：北京工业大学建筑与
　　　　　城市规划学院

2199

设计说明：
1. 主题思考
(1) 长寿建筑，全程低碳，空间利用率高
(2) 多种家庭人口组合要求（多代共享）
(3) 充分利用自然条件（太阳能、风能）
2. 地域性思考
(1) 严峻的冬季低温——保温、采暖
(2) 强劲的冬季西北风——阻挡、利用
(3) 较弱的夏季西南风——加强、组织
3. 太阳能技术应用
(1) 被动太阳能利用：太阳墙、太阳能集热模块、太阳房
(2) 主动太阳能利用：太阳能热水器、太阳能光伏电池板
(3) 风能：太阳能烟囱、地下热能交换通道、风力发电

Economic and Technic Index
Overall land area   1.1hm²
Overall building area   22000m²
Building density   21.3%
Number of families   240
Floor area ratio   2.1
Greening rate   58.9%
Ground parking rate   10%

SOLAR WIND

Master plan 1:500

LOCATION ANALYSIS

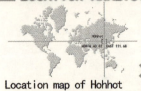
Location map of Hohhot

Solar energy resources distribution

More resource region (Solar radiation amount 130-150 kcal/(cm²·year)

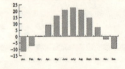

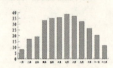
Temperature Analysis (°C)

Solar cell

1. About subject
(1) Long life building, low carbon, efficient use of space
(2) Variable living unit
(3) According to natural condition
2. About local characteristic
(1) Heat-preserving, heat-absorbing
(2) Wind1 (WN) blocking, using
(3) Wind2 (WS) strengthen, organizing
3. Solar energy technology
(1) Passive utilization: solar wall, collector module, solar house
(2) Active utilization: solar water heater, solar cell
(3) Wind energy: solar chimney, geoheat exchange channel, wind power windmill

SITE ANALYSIS

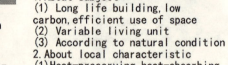

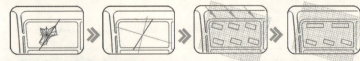

SITE PLAN ANALYSIS

Central courtyard

Traditional courtyard

The whole structure absorb many heat in cold month and it is cool in summer times. It means that the site planning is very active in solar energy using.

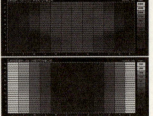

Functional partition analysis

Traffic streamline analysis

CARBON RELEASE IN DAILY LIFE

　　There are two main activities which release most amount of carbon in our daily life, travelling by car and using aircondition.
　　First, encourage people walk or by bike but not by bus. Second, make efficient use of the solar resource to keep the temperature within a comfortable area, there we mainly use solar energy and help with the wind energy.

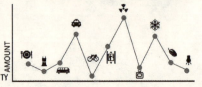

ARCHITECTURE

ELEVATION

Vertical axis wind turbine
Horizontal axis wind turbine

South Elevation (west)

West Elevation

Typical Floor Plan

First Floor Plan

PLAN

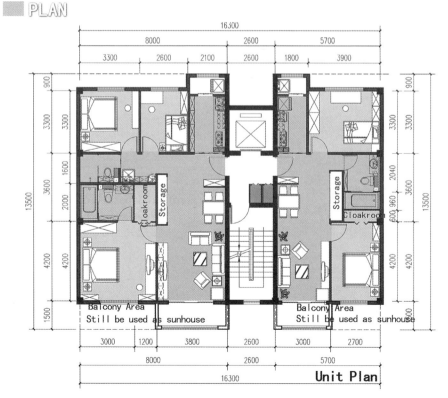

Unit Plan

Alterable Room

- Living Room
- Bedroom
- Bathroom
- Balcony
- Another choice of balcony

Flexible plane creates different habitable spaces, in order to meet different needs.

TECHNOLOGY

SOLAR WIND

WIND

Section 1-1 1:200   Section 2-2 1:200

Wind Organization

In planning

This distribution can against the cold wind in winter. In summer, through-draught go into the site, and be accelerated by the narrow distance between two gable walls.

In Architecture Design

Solar house (balcony)
Solar chimney
Geoheat exchange channel

WINTER — Warm air — Solar house closed
SUMMER — Through-draught — Solar house opened

All Technic applications

- Double layer floor
- Double layer wall
- Solar Chimney (Stair hall)
- Break bridge window
- Tridimensional virescence
- Solar wall
- Solar house
- Solar collector modules

Solar Energy Utilization

Wind Power Windmill

Vertical axis wind turbine | Horizontal axis wind turbine

Solar Chimney

Solar collector modules

Roller shutter up — Hole in the wall — Ceiling with holes
Thermal disspation
Damper opened

Roller shutter down — Hole in the wall — Ceiling with holes
Thermal disspation
Damper closed

Break Bridge Window Frame

Side-hung window has more advantages in gas tightness and water tightness, at the same time, it will hamper the noise outside.
The heat of each side will be preserved, forthermore, temperature condition is more stable than the room which use traditional window frame.

Solar Wall Air System

Air cavity
Hole
Air
Solar collector board
Air is warmed up

Operating principle

- Auto roller Damper
- Heat-absorbing board
- xps plate
- Glass cover
- Adjustable air interlayer
- Concrete heat-absorbing wall
- Temperature controlled damper

Construction

TECHNOLOGY
STRUCTURE COMBINATION

SOLAR WIND 4

Steel structure

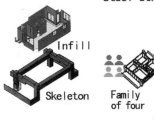
Infill / Skeleton

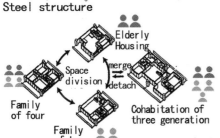
Space division — Elderly Housing / Family of four / Family of two / Cohabitation of three generation (merge/detach)

Steel structure: Short period, less wet operations. And it needs less building materials, which is recycled, supports SI system, and structural elements is small. The space that can be used is larger.

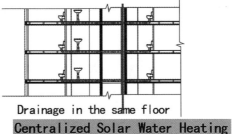
Drainage in the same floor

SI system with spatial flexibility

SI system: Determine basic column grid size based on spatial variation, combined with the arrangement of beams to form the S (skeleton). Floor, wall and various pipeline equipment are filled in the structure of the framework, to form I (infill).

Centralized Solar Water Heating

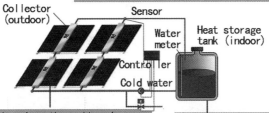
Collector (outdoor) / Sensor / Water meter / Heat storage tank (indoor) / Controller / Cold water

Placed on the wall under the window in the south wall | Placed in the balcony or bathroom

Green Planted Roof
- Vegetation
- Growing medium
- Filter membrane
- Drainage layer
- Waterproof root repellant layer
- Support panel
- Thermal insulation
- Vapour control layer
- Structural support

Heat transfer efficiency
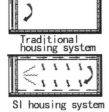
Traditional housing system / SI housing system

Horizontal shading

- Summer solstice
- Spring (Autumn) equinox
- Winter solstice

Planting pool

Water Recycled
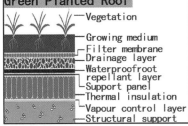
Solar cell / Rain water / Recycled water / Electricity / Water pump / Washing cars / Waste water / Flushing toilet / Irrigation

Solar Parking
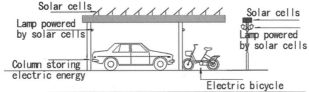
Solar cells / Lamp powered by solar cells / Column storing electric energy / Electric bicycle

Garbage Collection System
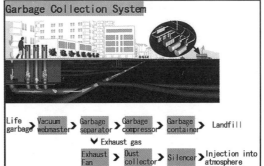
Life garbage → Vacuum webmaster → Garbage separator → Garbage compressor → Garbage container → Landfill
↓ Exhaust gas
Exhaust Fan → Dust collector → Silencer → Injection into atmosphere

Arrangement of Plants
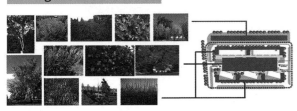

优秀奖
Honorable Mention Prize

项目名称：灿阳千里
One Thousand Splendid Sun
作　者：吴　巍、邢　茜
参赛单位：天津大学建筑学院

灿阳千里
ONE THOUSAND SPLENDID SUN
阳光与低碳生活 Low-carbon Life with Sunshine

设计说明：

方案用地位于内蒙古呼和浩特市新区，用地北临城市干道。规划用地范围总面积11000㎡，实施方案建设用地范围面积3890㎡。考虑用地范围内已建有自来水、排水、雨水、天然气、供电及电信系统，该方案着重从被动式太阳能建筑入手，辅助以主动式太阳能的技术进行设计，试图通过一种有效的设计，减少建筑物能耗，切实提高人们对建筑节能的认识，同时提高人们生活的品质。

Design report:

The program located in Inner Mongolia, Hohhot City District, north to the city roads. The total planned area is 11,000 square meters, and the total area of the project is 3,890 square meters. Considered within the land there has been built with tap water, drainage, water, gas, electricity and telecommunication systems, the project focuses on starting from the passive solar building, secondary to active solar technologies. We try to find an efficient way to reduce building energy consumption and improve people's understanding of building energy efficiency, in the same time increasing the quality of people's lives.

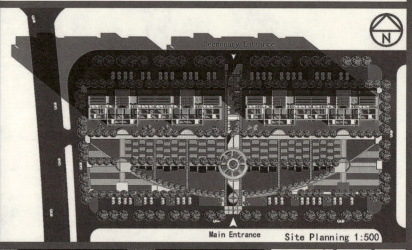
Site Planning 1:500

Stereographic Diagram

Psychrometric Chart

Location

The Main Technical and Economic Indexes

- Total Land Area : 1.1 h㎡
- Overall Floorage : 18581.6 ㎡
- BUilding Density : 16.2 %
- Total Households : 176
- Floor Area Ratio : 1.7
- Greening Rate : 31.5 %
- Ground Parking Ration : 33 %

Bird's-eye View

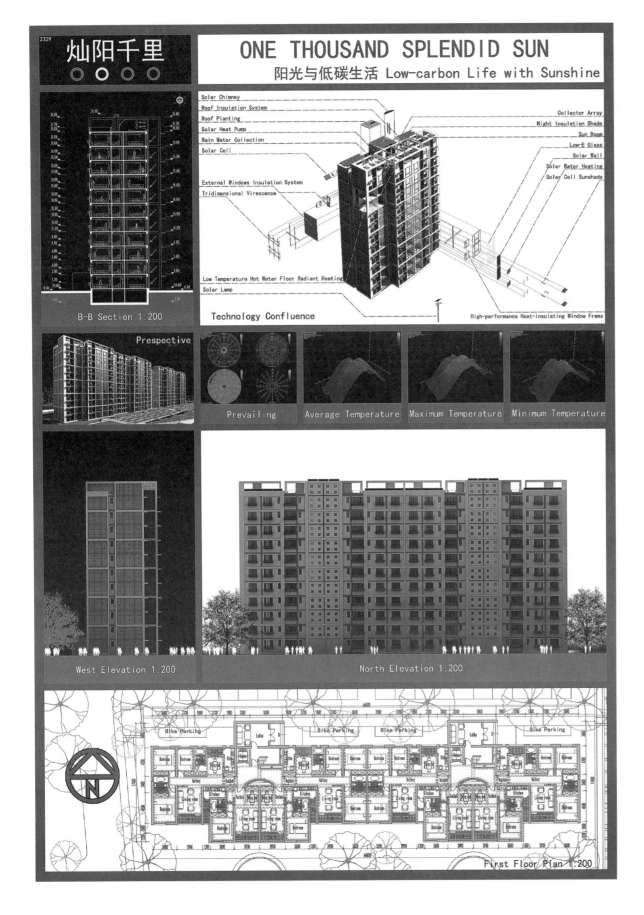

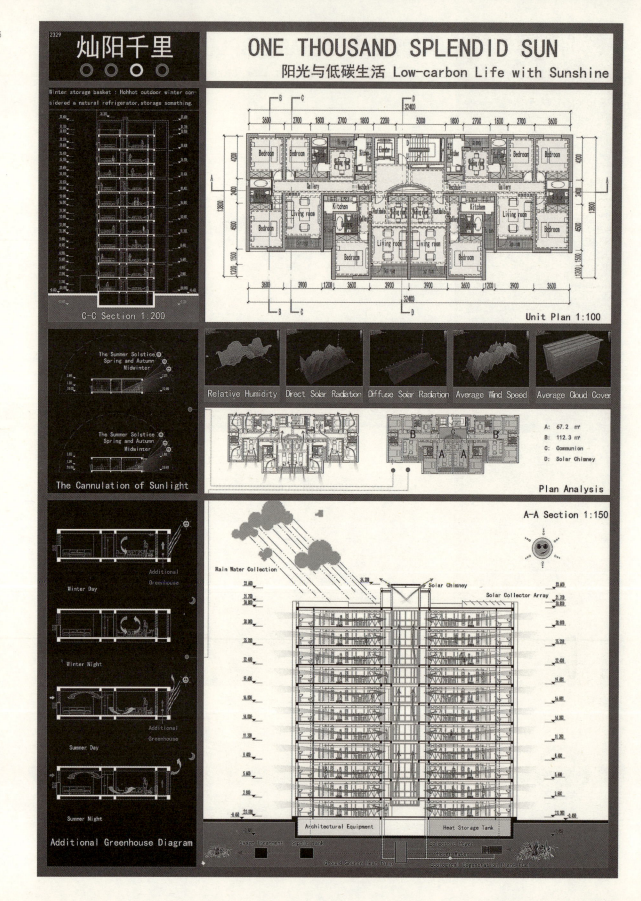

灿阳千里
ONE THOUSAND SPLENDID SUN
阳光与低碳生活 Low-carbon Life with Sunshine

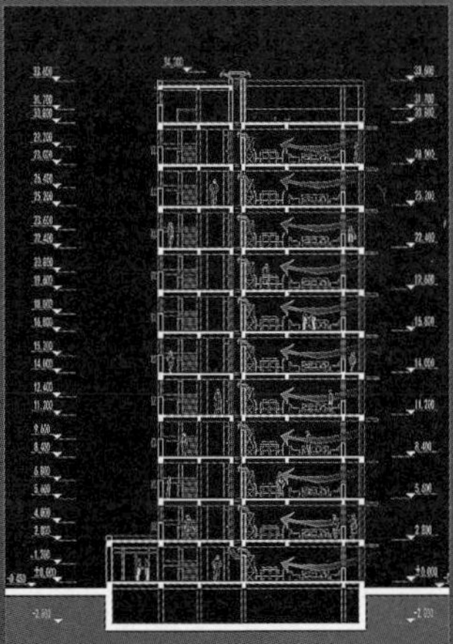

D-D Section 1:200

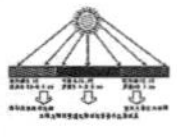

	Active Solar Energy Utilization	Solar Cell
		Solar Heat Pump
		Solar Collector
		Solar Water Heating
		Solar Battery Sun-shading System
Low-carbon Life	Passive Solar Energy Utilization	Additional Greenhouse
		Solar Wall
		Storage Hot Water Wall
		Low Temperature Hot Water Floor Radiant Heating
		Solar Chimney
		Phase Change Solar System
		Heat Storage Tank
	Architecture Energy Efficiency	Nature Circulation System
		Auxiliary Thermal Source - Ecological Cogeneration Plant Fuel
		External Windows Insulation System
		Area Ration of Window to Wall
		Roof Insulation System
		Roof Planting
		Rain Water Collection

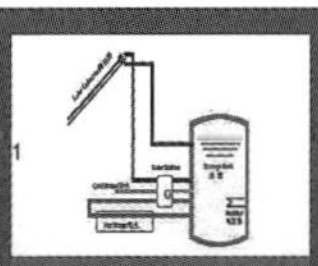

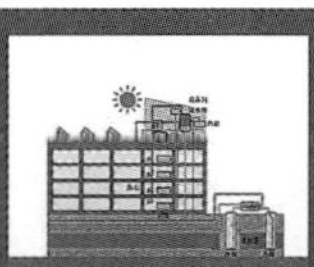

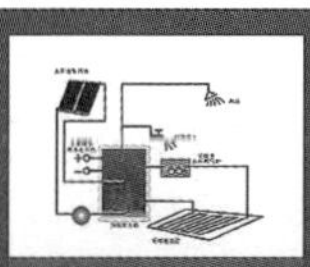

Perspective Drawing

优秀奖
Honorable Mention Prize

项目名称：新竹
　　　　　The New "Bamboo"
作　　者：彭 卓、杜 萌、胡松玮、
　　　　　潘 崟
参赛单位：重庆大学建筑城规学院

新竹 [吴江市低碳宜居住宅设计]

设计说明

　　设计首先从传统建筑入手，通过对竹子式住宅的探索，结合现在人们对居住生活的要求，采取了对传统建筑的重构，形成了按进深方向，通过天井组织，由公共逐步到私密的形态。结合当地城市肌理，形成联排的布局形式，呼应城市的建筑格局，节约建筑用地的同时丰富了内部的环境，中间恢复了传统水系，不仅丰富了景观，而且形成了良好的微气候。冷巷的布置可以有效带走夏季空调造成的热量，通过天井的排风，形成适宜的温度。局部的底层架空以及商业门面的加入，丰富了平时缺乏的交往空间。这种一房一水的居住模式在设计中延续，通过局部改进整体考虑的模式，节约资源，融入当地环境，达到低碳宜居的目的。

Design report

This design would focus on the traditional architecture, by exploring the bamboo style house and people's demand to living situation, we would start to recreate traditional architecture, which would transform the former ones into new architecture which features constructed in the direction of spatial depth, organized by courtyard, has a form which could gradually define spaces form private to public. According with the local architectural composition, the layout of the architecture would be set in rows, this would result in valuing building land and make the inner space more abundant, the recovery of traditional water in the middle site of building not only enrich the landscape but also create good micro-climate.The arrangement of cooling alley could bring away heat caused by air conditioner in summer in high efficiency city ventilation through the courtyard could form a suitable temperature. The part of the bottom and water living model continue modification in design, through improvement in local parts, the overall consideration model, conservation of resource which benefits local environment, to achieve the purpose of low-carbon livable buildings.

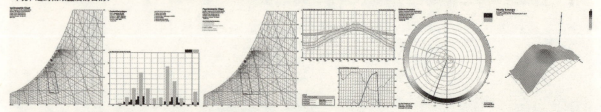

THE NEW "BAMBOO" [THE LOW CARBON AND LIVABLE RESIDENT DESIGN]

Climate and techniques analysis

1ST

新竹 [吴江市低碳宜居住宅设计]

2011 台达杯国际太阳能建筑设计竞赛获奖作品集

Outer gallery affecting lighting
Small patio affecting lighting
No light into inner room

Problems in traditional bamoo house

By utilizing the serve and be serve space theory of Luis Kang to cure the shortage of traditional bamboo style building, we put two living room in the south section which has the best sunshine and set a courtyard with appropriate dimension close by, then overhead the ground floor in sake of guide the airflow, these measurements would make the building have better perfomance in lighting and ventilating.

FIRST FLOOR PLAN 1:150

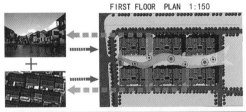

Main landscape axis
Main landscape node
Minor landscape node
Waterscape

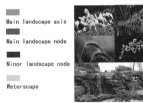

The material utilized by jiangnan traditional buildings and soil, stone, calcareousness, oyster dust, brick, tile, bamboos, bulrush, rice, grass, Chinese wood oil, paint and so on, these materials mentioned above are all natural, and shows no harm to natural environment.

SITE PLAN 1:500

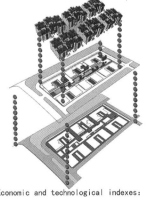

Economic and technological indexes:
Total construction area: 9023.25m²
Afforestation: 25.3%
Total households: 72
Building density: 56.3%
Land area: 3887.5m²

THE NEW "BAMBOO" [THE LOW CARBON AND LIVABLE RESIDENT DESIGN]

2nd

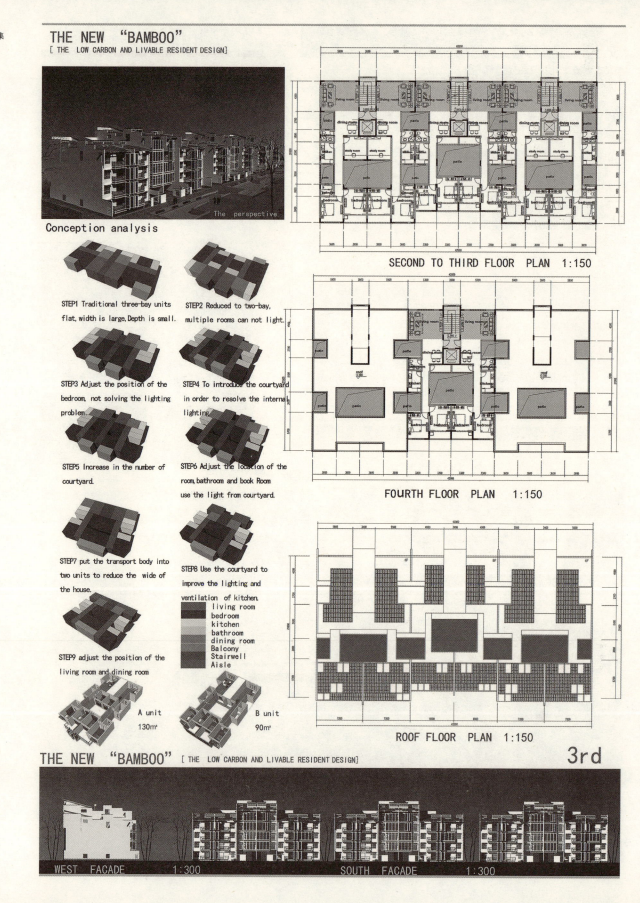

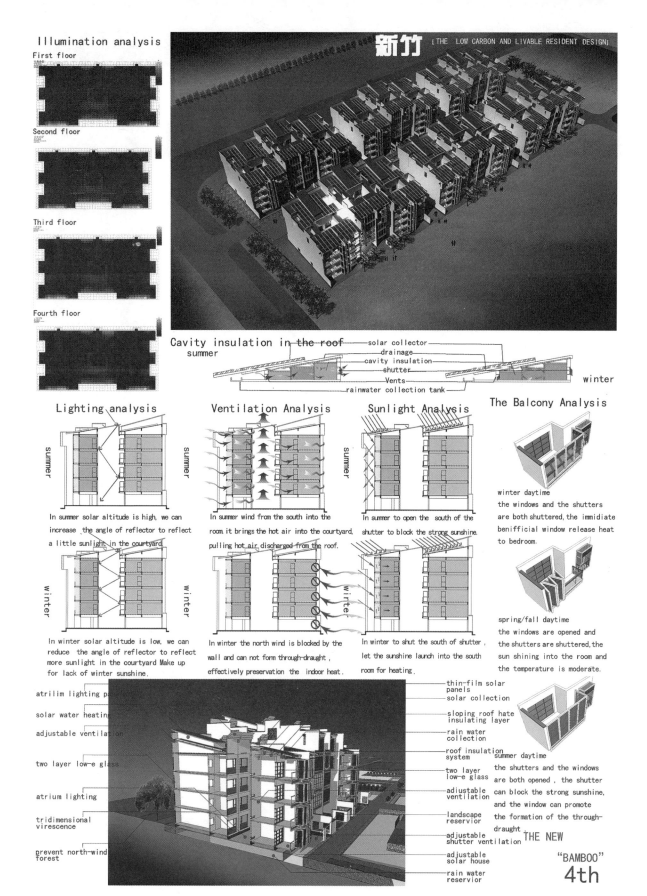

优秀奖
Honorable Mention Prize

项目名称：把阳光带回家
　　　　　Bring the Sunshine Home
作　　者：魏琪琳、陈露、张旭、
　　　　　李浩
参赛单位：重庆大学建筑城规学院、
　　　　　香港中文大学

BRING THE SUNSHINE HOME
把阳光带回家 01

Bird's-eye view

设计说明：本案地处呼和浩特，冬冷夏温。以两代居为着重点打造适应北方且满足老年人身心需求的太阳能住宅。探讨了高龄人群卧室整体引入阳光的可能性，结合凸窗设计太阳光多次反射装置，将阳光洒向了整个空间。同时加入了太阳房采暖、太阳能发电、太阳能热水采暖循环、太阳光光纤导入技术，并应用了多种绿色节能环保措施打造宜居绿色家园真正实现阳光、自然、生态。

Design illustration:

This case locates in the city of Huhhot. The weather there is extremely cold in winter and warm in summer. The design aims at creating a Solar Residential for the weather condition in northern China and maintaining the requirements of the olds in both physical and psychological aspects. In the design, we are searching for the possibility to bring the sunlight to every corner of the bedrooms for the olds. And we made it by using multiple sunlight reflection devices with certain bay-window design in the bedroom. Meanwhile, solar house heating, solar power, solar water heating loop, sunlight-import optic fiber technology as well as multiple green energy saving measures are introduced to this project for creating a green livable home. In this way, our aim for sunlight, nature ecology can be achieve.

The shutter could all be folded above to let the light get trough the window derectly.

Adjust the angle of the shutter to let the light get in horizontal direction or to the reflector.

Adjust the angle of the shutter to let the light get in to the reflector.

The arc reflector is to make the light diffusion.

Reflector Analysis

= the arc reflector + the reflective shutter + the heat insulation window + the bay-window

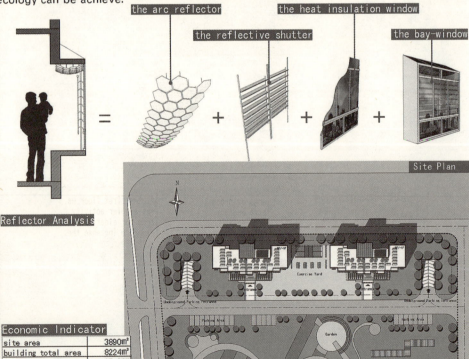

Site Plan

Economic Indicator	
site area	3890㎡
building total area	8224㎡
building density	27.90%
number of household	84
floor area ratio	2.1
greening rate	39.40%
ground parking rate	12%

SOLAR ENERGY ARCHITECTURE DESIGN

BRING THE SUNSHINE HOME

Summer: curtain prevents light in. open the intakes to air the room. water wall isolate heat.

Winter: solar house collects heat and help make air flow in the room.

Sunshine collector

It is inevitable for plate high-rise dwelling houses in northern China. In order to meet the requirement for daily-live energy saving, it needs to improve the design of solar energy collect devices and save the energy by using optical fiber to deliver the sunlight to certain rooms.

- Condenser components (Convex lens)
- Fixtures
- Sensors (Sun position)
- Sunshine collecter
- Base

Convex lens → Cut convex to hexagon → Convex hexagon collector → Assembled into a sunshine → Assembled

Elderly Residential

Amount of sunlight can improve sleep quality.
Amount of sunlight can anti-cold.
Amount of sunlight can extend people's life.

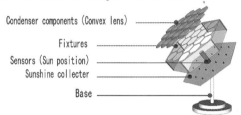

Bring the sunshine to bedroom, we can enjoy sunlight and healthy indoor, it can be a good way of healthy and livable.

In our two generation residential, we have four house types, in order to adaptation the modern family which has two generations, we can combined the house type B and C to type D, it will be independent and fusion. Then have a big handing.

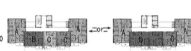

Analysis of sunshine import

through the sunshine collector and light tank, sunlight is sented to the certain room, then achieve the target.

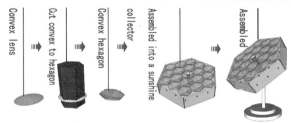

- sunshine collector
- fiber pipe
- sunshine collector
- light tank

The characteristics of the old people's residential:
large bathroom,
large bedroom,
large solar house,
compact living room,
spacious walkways,
south bedroom.

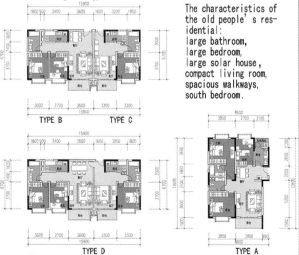

TYPE B TYPE C
TYPE D TYPE A

Perspective profile

At the first floor, we set a solar activity room for elderly, in order to increase the contact of old people, we set a hanging garden every interval level.

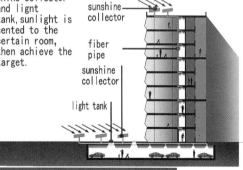

- aisle
- solar house
- hanging garden
- solar house
- hanging garden

SOLAR ENERGY ARCHITECTURE DESIGN

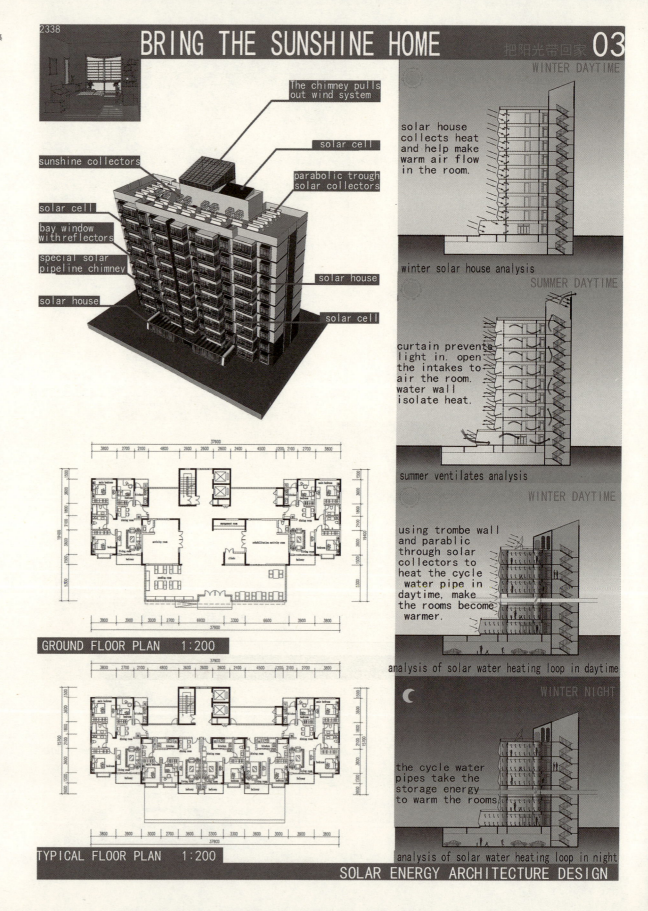

BRING THE SUNSHINE HOME — 04

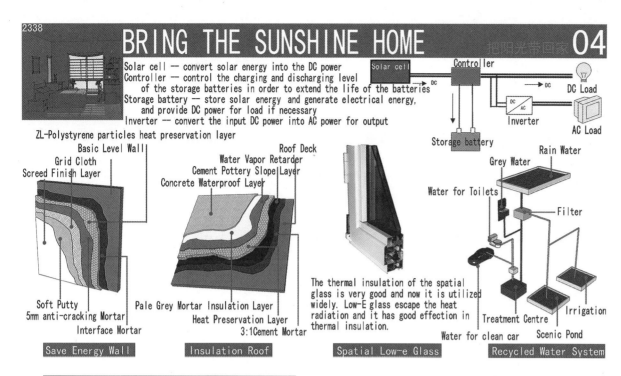

Solar cell — convert solar energy into the DC power
Controller — control the charging and discharging level of the storage batteries in order to extend the life of the batteries
Storage battery — store solar energy and generate electrical energy, and provide DC power for load if necessary
Inverter — convert the input DC power into AC power for output

Save Energy Wall
- ZL-Polystyrene particles heat preservation layer
- Basic Level Wall
- Grid Cloth
- Screed Finish Layer
- Soft Putty
- 5mm anti-cracking Mortar
- Interface Mortar

Insulation Roof
- Roof Deck
- Water Vapor Retarder
- Cement Pottery Slope Layer
- Concrete Waterproof Layer
- Pale Grey Mortar Insulation Layer
- Heat Preservation Layer
- 3:1 Cement Mortar

Spatial Low-e Glass
The thermal insulation of the spatial glass is very good and now it is utilized widely. Low-E glass escape the heat radiation and it has good effection in thermal insulation.

Recycled Water System

Analysis Of The Chimney Pulls Out Wind System

In winter, close the Window and black wall will absorb solar energy which can heat the air of the entire staircase inside.

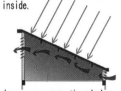
In summer, open the window and black wall will absorb solar energy which can pull out wind because of wind pressure and hot-pressing.

Section | Facade

SOLAR ENERGY ARCHITECTURE DESIGN

优秀奖
Honorable Mention Prize

项目名称：向阳门第
　　　　　Sun Mile
作　　者：李　龙、韩梦涛、王君益
参赛单位：华中科技大学建筑与城市规划学院

ID:2428　Low-carbon Life with Sunshine-Solar Building Design

Sun Mile residential "向阳门第" MY HOME. MY FUTURE
绿色节能居住区设计

NOW, What can we do?

A Main topic:
A. Scheme consideration
(1) Sun and residence. Lighting is an essential necessity for living environment. The design not only manages to improve lighting but also saves resource by using solar energy.
(2) Use of rain water and wind power. Wujiang district is located in southern coastline with adequate rain falls and sea wind, which afford with us the possibilities of utilizing rain water and wind power.

B. Locality
The building centers around courtyard with symmetrical layout. Jiangnan traditional architectural elements are integrated with the shape of the building.

C. Energy savings
(1) Solar energy: Solar electrical generation facilities, water recycling system, heating-storing walls, potable shutter, methance.
(2) Rain water: Rain water collecting system, rain water purifying and recycling facility, wetland landscape, park characteristics.
(3) Wind power: wind electrical generation facilities, wind power landscape.

设计说明
A 主题思考：
(1) 阳光与居住——良好的采光是人们对居住环境最基本的要求，同时综合利用太阳能以节能。
(2) 雨水和风能的利用——吴江地区地处南方沿海城市，降雨量充沛，海风充足，可充分利用雨水和风能。
B 地域性思考：
以园林式庭院为核心，对称布局，将江南民居建筑立面元素融入建筑造型之中。
C 节能技术综合利用思考：
(1) 光能——太阳能发电，生活热水循环系统，蓄热墙体，可调节式百叶，沼气利用等。
(2) 雨水——雨水收集系统，雨水净化再利用，湿地景观，园林特色。
(3) 风能——风能发电，风能景观。

Low-carbon Life with Sunshine-Solar Building Design

Sun Mile residential "向阳门第" MY HOME · MY FUTURE
绿色节能居住区设计

NOW, What we can do:

Green wall design

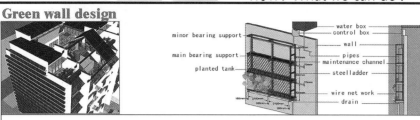

Reflector and Solar panels

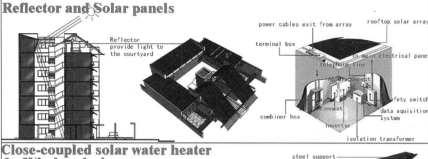

Close-coupled solar water heater & Window design

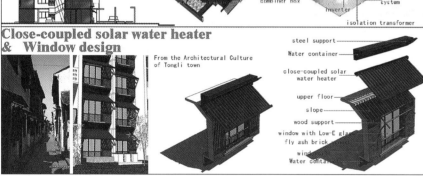

Area position

Climate: Wujiang, located in Jiangsu Province, faces toward East China Sea in east and Taihu Lake in west. It has north subtropical zone damp monsoon climate, attributed to which it is humid and rainy with high temperature in summer while dry and cold in winter. Monsoon and four distinct seasons are also clearly here.

Layout Analysis | **Plane Analysis**

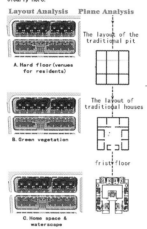

A. Hard floor (venues for residents) — The layout of the traditional pit

B. Green vegetation — The layout of traditional houses

C. Home space & waterscape — first floor

向阳门第   SUN MILE residential

Household Designing

household plan 1:50 ■ subordinate living space □ primary living space

daylight factor of the original programme

daylight factor of the improved programme

During the daylight factor analysis of the two programme, it is showed clearly that the improved one gets more illumination which is attributed to the atrium added. It also can help enhance ventilation.

- - - alignment line
→ illumination route

compound mode of the household and route for light

3D model of 90m² residential
3D model of 130m² residential

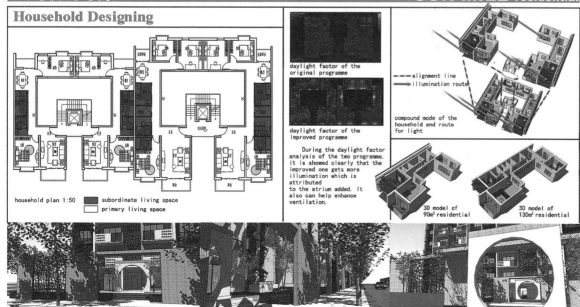

ID:2428 Low-carbon Life with Sunshine-Solar Building Design

Sun Mile residential "向阳门第" MY HOME. MY FUTURE
绿色节能居住区设计

NOW, What we can do:

Ventilation & Filtration

Bottom elevated can enhance ventilation.

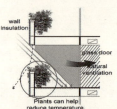
Open atrium can enhance funnel draft.

wall insulation / glass door / natural ventilation
Plants can help reduce temperature.

Detail of Double-layers' Roof Ventilation System

shutter | auxiliary equipment | double-layers' roof | ventilation system of double-layers' roof | SUMMER | WINTER

With double layers' roof, during summer, we open the north window in the roof, let the sun heat the air between the roof and make the warm air be ventilated from inside. Assistant ventilation system can be used when necessary.
During winter, with the window closed, the room will be heated with the circulation between the warm air within the roof and the cold air indoor.

First Floor Plan with Landscape

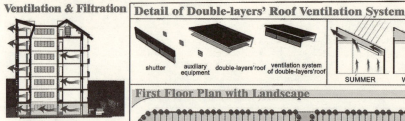

wind-light complementary

Analysis of Annual Tempreture Indoor

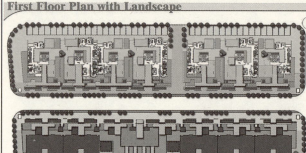

diagram of annual tempreture indoor in **OLD** 130m² residential
+atrium +bottom overhead
diagram of annual tempreture indoor in **NEW** 130m² residential

diagram of annual tempreture indoor in **OLD** 130m² residential
+atrium +bottom overhead
diagram of annual temperature indoor in **NEW** 130m² residential

■ above 35 ℃
□ 29~35 ℃
□ 20~29 ℃
□ 16~22 ℃
■ below 16 ℃

These diagrams show in the whole year, how the tempreture attributes in every room of the residential only by ventilation without air-conditioner. There are differences between OLD & NEW residentials. It is caculated by DeST-h.

向阳门第 SUN MILE residential

PROGRAM GENERATION WITH VENTILATION

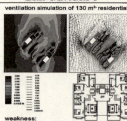
Wind passes through the site without buildings straightly.

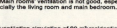
The layout of our buildings can guide wind into the block in summer and prevent wind blowing in it in winter.

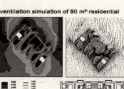
The atrium would lead natural wind inside, lowering the temperature.
The funnel draft of the atrium manage to improve ventilation of the building in summer.

← predominant wind in summer
← predominant wind in winter

The design considers natural ventilation in terms of site, volume and space in order to lower the temperature in summer and save energy of air-conditioners.

Ventilation Simulation(CFD)

OLD CONCEPT

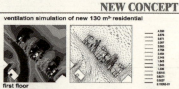
ventilation simulation of 130 m² residential

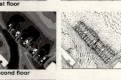
weakness:
Main rooms' ventilation is not good, especially the living room and main bedroom.

ventilation simulation of 90 m² residential

weakness:
Main bedroom' ventilation is not good.

NEW CONCEPT

ventilation simulation of new 130 m² residential

first floor

second floor

sixth floor

section

new 130 m² residential

new 90 m² residential

advantage:
Main rooms' ventilation is better, especially the living room and main bedroom. Because of the atrium, the vertical ventilation is better. And the residential can get more illumination.

Low-carbon Life with Sunshine—Solar Building Design

ID:2428 | 04

Sun Mile residential "向阳门第" MY HOME · MY FUTURE
绿色节能居住区设计

水 — Rain water collect | Wetland landscape | Water floor

NOW, What we can do:

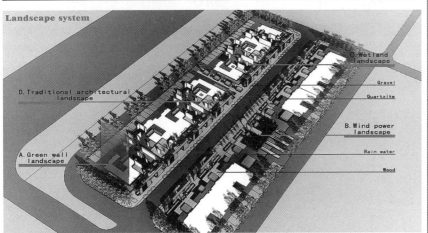

Landscape system
- A. Green wall landscape
- B. Wind power landscape
- C. Wetland landscape
- D. Traditional architectural landscape
- Gravel
- Quartzite
- Rain water
- Wood

Detail of wetland landscape
Perspective: A / B / C / D

Energy saving system

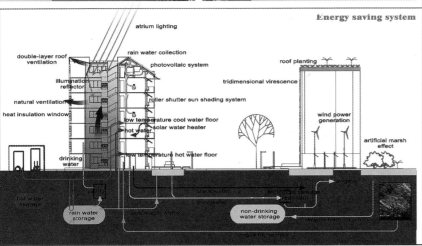

- atrium lighting
- double-layer roof ventilation
- rain water collection
- photovoltaic system
- roof planting
- illumination reflector
- tridimensional virescence
- natural ventilation
- roller shutter sun shading system
- heat insulation window
- low temperature cool water floor
- wind power generation
- hot water
- solar water heater
- drinking water
- low temperature hot water floor
- artificial marsh effect
- hot water storage
- rain water storage
- landscape water
- black water
- gray water
- biological treatment
- non-drinking water storage
- irrigate
- biogas for kitchen

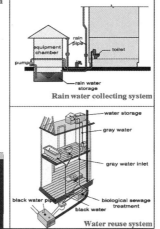

Rain water collecting system

Water reuse system

向阳门第   SUN MILE residential

Perspective

优秀奖
Honorable Mention Prize

项目名称：阳光、水、人家
　　　　　Sunshine, Water, Home
作　　者：陈稀
参赛单位：河北工业大学

SUNSHINE, WATER, HOME

DESIGN NOTES

The design respected to the traditional residential, combined with solar energy technology and created constructed wetland that can purificate rainfall.

建筑设计与当地传统民居和生活方式相结合，利用太阳能通风技术、主动和被动太阳能采暖技术以及太阳能光伏技术，结合当地实际情况制造人工湿地收集并利用植物净化雨水。将传统民居风格、太阳能和雨水回收利用相结合。

TECHNICAL AND ECONOMIC INDICATORS

Total land area : 1.1hm²
Total building area :18822m²
Building density :28.52%
Total households :180
Floor area ratio :1.71
Greening ratio :38.12%
Parking ratio :26.11%

CLIMATE ANALYSIS

Degree hours
- heating
- cooling
- solar

Rainfall and Relative humidity

Temperature
- Maximum
- Average
- Minimum

Daily solar radiation

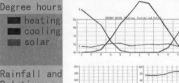

Best orientation
Location:Wujiang,CHN
Underheated Stress:1160
Overheated Stress:209
Comprimise:175

Wind analysis
Location:Wujiang,CHN
(31.1°,120.6°)
Date:1st January-31st December
Time:00:00-24:00

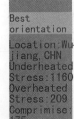
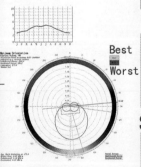
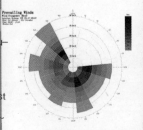

MASTER PLAN 1:500

rainwater recycling | water seepage ground | city road
solar photovoltaic roof | constructed wetland | plants shading | sunspace roof

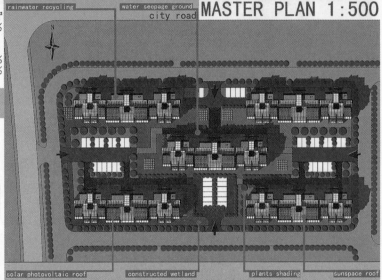

VENTILATION·SUNSHINE DURIATION

summer | winter | sunshine duriation

SOUTH ELEVATION/EAST ELEVATION 1:200

DESIGN OF WUJIANG APARTMENT

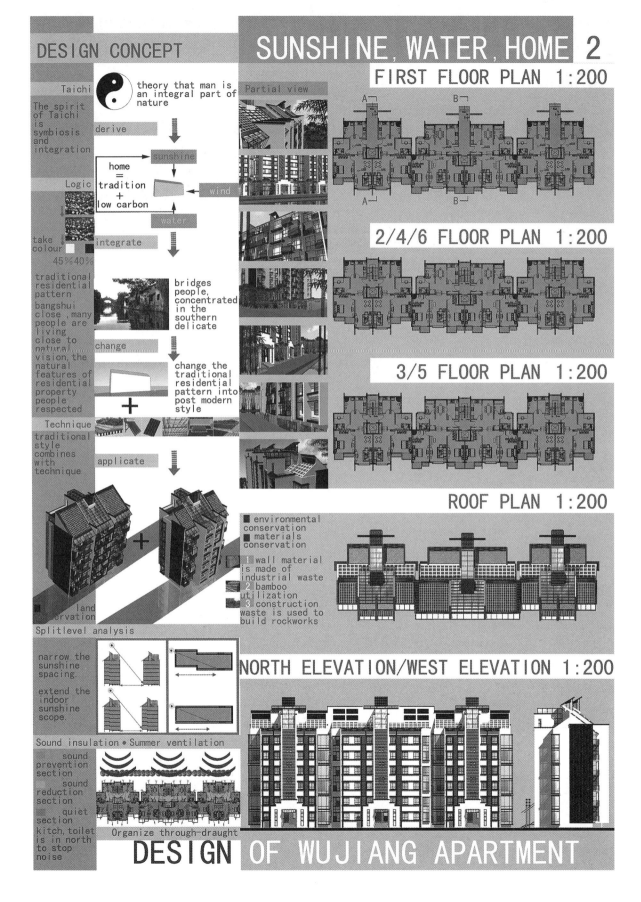

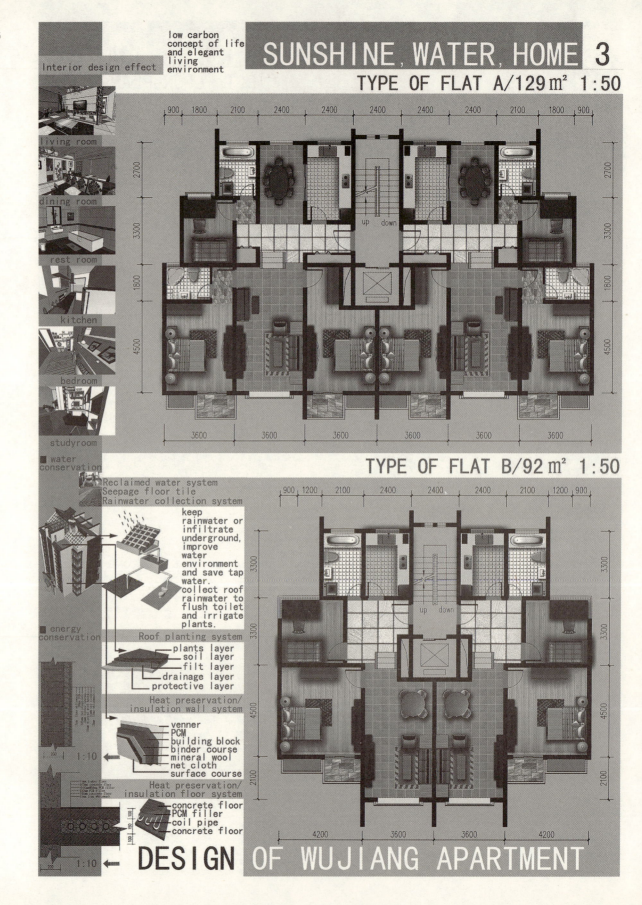

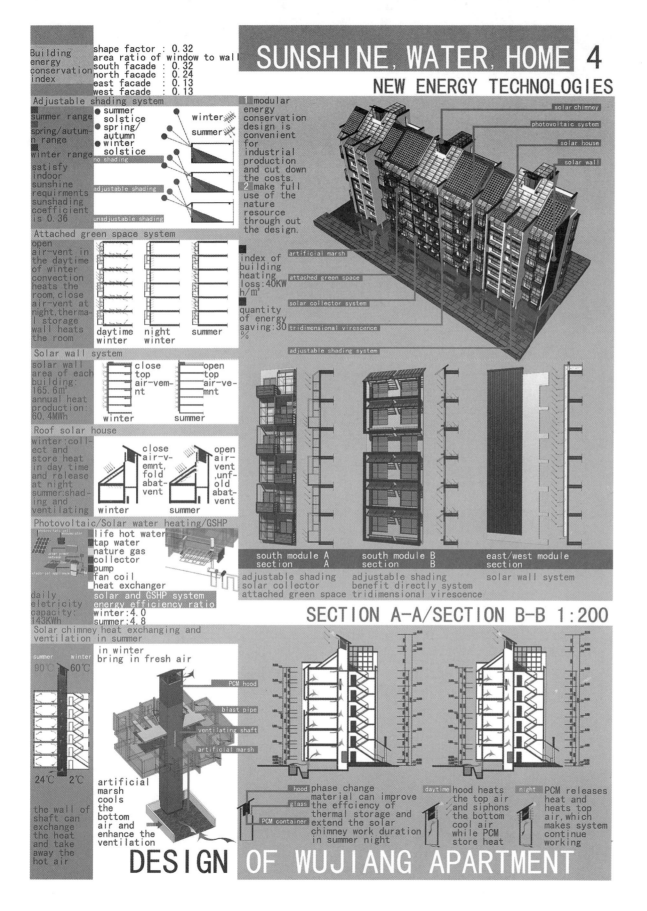

SUNNY VALLEY 阳光谷
sunny apartments

2556

优秀奖
Honorable Mention Prize

项目名称：阳光谷
　　　　　Sunny Valley
作　　者：Xavier LAGURGUE、李 芬
参赛单位：XLGD 建筑师事务所

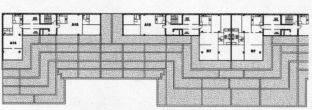
5th floor / 300e

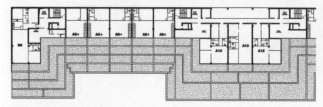
4th floor / 300e

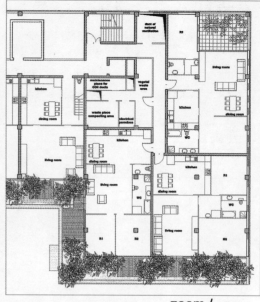
zoom / 100e

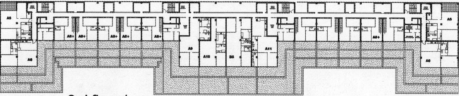
3rd floor / 300e

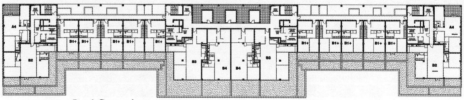
2nd floor / 300e

1st floor / 300e

east and west elevation / 200e

SUNNY VALLEY 阳光谷
engineering principles

2556

Carbon cycle

- filter substrate
- felt breathable
- floor on pads
- pulsed CO₂ fan

Boilers produce CO₂ which is injected inside the terrace false floor. CO₂ is filtered by the substrate and helps plants' growth during photosynthesis.

Phytoepuration

- canopy of photovoltaïc capture
- treated water
- rain water
- purification plants
- to methane unit
- used water
- drained infiltration

Used water coming from sink or toilets is filtered by phytoepuration at the basement of the building, under the canopy capture which need two square meters by person. The vegetation product feed for the methane unit. The purified water is ascended too the floors with a pump and used for flush and watering.

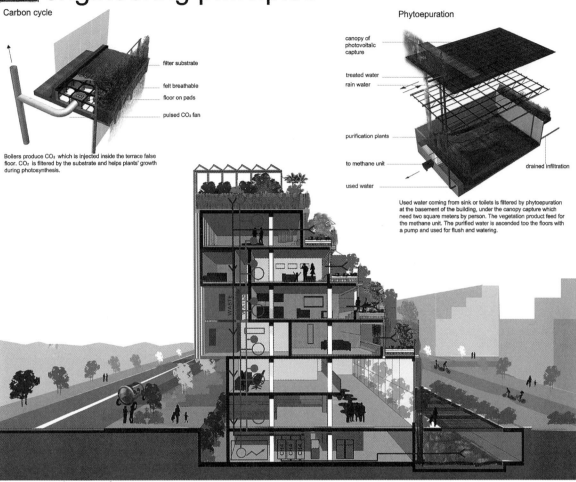

perspective cut / 200e

Water cycle

- lift pump
- watering
- waterfall of rain
- perspiration - evaporation
- canopy of photovoltaic capture
- phragmites, reeds, iris, mints
- exit of traited water
- arrived : wastewater and waive water
- drained infiltration

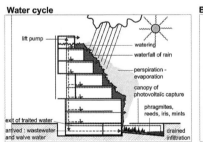

Biomass production

- plants, biodiversity support
- organic wastes
- phyto-purification plants
- methane unit

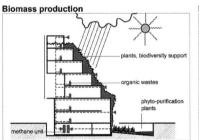

Mutualisation of energy

- hybride car share (gasoline + electricity)
- electricity produced by methanisation
- electricity produced by cogeneration

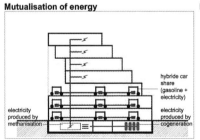

Hybride cars

- photovoltaïc capture
- auxiliary fuel : gasoline
- electric battery

The hybrid cars are in free access at the building parking. They are the electricity reserved, specialy at the peak hours during the morning and evening. They are feed during the night with the cogeneration unit, and by day with the gasoline and photovoltaïc capture.

Energy production

- methane unit
- methane's boiler
- waste's boiler
- cogeneration unit
- hot water tanks

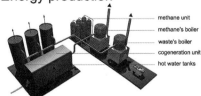

Wastes are a supply. They are selective sorting by each dweller. Only plastics and metal are removed. The organic waste feed the methane unit, the non-organic waste are burned in a boiler to feed the cogeneration.

SUNNY VALLEY 阳光谷
an urban district

2556

building view

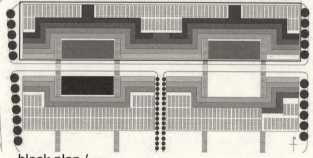

block plan / 500e

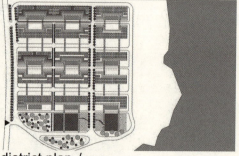

district plan / 2000e

The project considers the three cycles of energy, carbon and water. It's based on:
- The symbiotic relation between vegetation and dwelling with a urban agriculture shaped in terraces affirmed as a characteristic of chinese landscape.
- the valorisation by cogeneration of organic and combustible waste.
- a new durable technology of electric car that stock the power for the building.

The building marks the beginning of a new plot system which is composed by a urban facad on the street side and a vegetal landscape at the heart of each plot. The planted terraces bring the souvenir of typical chinese rustic sites and sunny valley in the city. The rain and watering creates streams and waterfalls in a theatrical scene which start from the upper floors. They are part of a landscape design identity proper to the TongLi site. The project is composed of housing and shops, guest rooms and offices to improve the fonctional mixity and reduce commuting. It's also a power plant and has recycling systems. The sun creates biomass by photosynthesis. The organic matter is transformed in methane which feed the cogeneration units producing hot water and electricity. The electricity is stocked during the night by electric cars. A photovoltaic system and a boiler for waste combustion complete the electricity production.

方案综合考虑了能源、碳和水的循环利用。
设计的原则建立在：
— 利用屋顶平台结合中国景观特色发展都市农业，研究人类居住模式与植被的共生关系。
— 利用生物质能和垃圾发电及水处理提升资源循环利用价值。
— 利用新型电动汽车等绿色新能源技术来进行能源储存和调配。
建筑的格局从选取城市一个岛状住房群的起端开始，通过中国内敛的居住哲学，闭合朝街的城市立面和朝内核心岛绿化面组合构成逐层退台的方形立体花园。
葱郁的植被平台如梯田般，给城市带来乡村景观特色，让人如驻足于阳光明媚的山谷的别样世界。雨水收集体系回灌屋顶花园的植物，潺潺的跌水从上层开始流向下层，穿插于具有同里古镇江南水乡黑白色系的建筑物，及五彩缤纷的布艺遮阳和绿野之间。
方案混合了商住、访客住所和办公等多种功能，目的是通过改善城市的交通组织，减少出行，和利用收集共建的余热回收，实现能效转换和调配，以提高能效利用。
光合作用产生生物量的原理，有机质转化甲烷燃烧产热，同时朝内花园阳立面的太阳能薄膜和地面的光伏发电体系，及屋顶的太阳能板，共同进行太阳能收集，实现三联供。能源的储存白天则由电动车发动机电池进行蓄能和调节，而在夜间保障能源的供给。

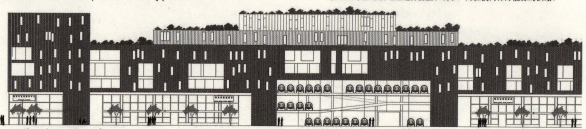

north elevation / 200e

SUNNY VALLEY 阳光谷
inside the valley

2556

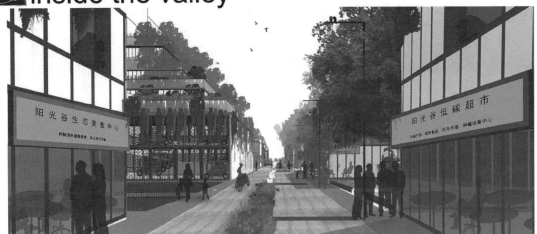

block view

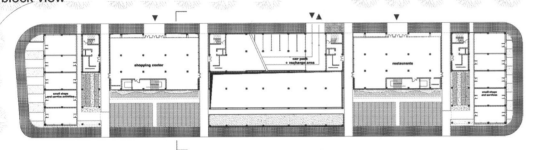

ground floor / 300e

basement floor / 500e

serial number	item	unit	remark
1	total site area	hm²	0,3965 (3 965 m²)
2	total floor area of the building	m²	19 589
3	building density	%	82
4	number of dwellings	unit	54
5	floor area ratio	%	494% = 4,95 per 1
6	greening rate	%	66
7	ground parking rate	%	1,5 per dwellings 15%

Hot water production

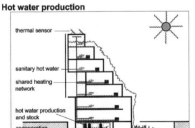

Thermal comfort
The winter comfort is ensured by:
- an airtightness skin (leak flow <0,6 m³/h),
- an auxiliary heating per hot water network from the cogeneration (C<15kwh/m²).
The summer comfort is ensured by:
- caduc plants from the terrace that give shadow during the summer and let the sun through during the winter.

Ventilation and quality of indoor air
A natural ventilation for the summer confort and hygiene based on the thermosiphon effect between the north and south facades.

Thermal comfort

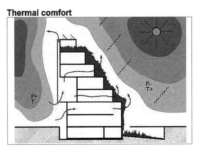

south elevation / 200e

优秀奖
Honorable Mention Prize

项目名称：忆江南
　　　　　The Memory of Jiangnan
作　　者：赵　林、刘文峰、冷　鑫、
　　　　　宗　伟、窦瑞琪、殷晓峰
参赛单位：山东建筑大学建筑城规学院

THE MEMORY OF JIANGNAN
SOLAR ENERGY ARCHITECTURE DESIGN

Wujiang is a city with a long history and has a plethora of scholars and talents. It is also the famous tourism city in the Yangtze River Delta, China, with its traditional culture and silkworm civilization, and the colorful historic sites and relics. Its Tongli Tuisi Garden - the listed World Culture Heritage, is undoubtedly a bright pearl in the Yangtze River Delta.

The concept of this project is based on the relationship between sunshine and daily life. By taken the present situation of the layout area and surrounding environment into consideration, the buildings are in coordination with the natural and man-made landscape, with the analysis of both the scenery elements and the weather conditions in the technical and cultural way. Benefiting from local building materials, we have taken advantages of traditional houses and have made a great progress. Thus the regional cultural has continued and the intimacy of traditional houses has been reserved.

Considering the aspects of technology we take the principle of environmental protection, energy conservation, and economic saving. We have quite an amount of technologies into a reasonable system, such as photovoltaic power generation, trombe wall, solar roller shutter, water recycle system, etc. In this method, we will lead to a comfortable life as the buildings fulfill the using function in the situation of low energy consumption.

设计说明 Design Description

设计构思以阳光与生活为中心展开，充分考虑基地现状与周边环境，准确把握景观要素与气象条件，合理结合先进技术与传统文脉，进行住宅建筑与景观环境的综合布置，采用当地建材结合当地传统民居意向进行设计演化，在对当地地域性文化延续的同时营造了民居般的亲切感。技术运用以环保、节能、经济为原则，将光伏发电、特龙幕墙、太阳能百叶、雨水收集等技术系统整合，使建筑在低能耗的条件下满足使用功能，营造舒适的居住环境。

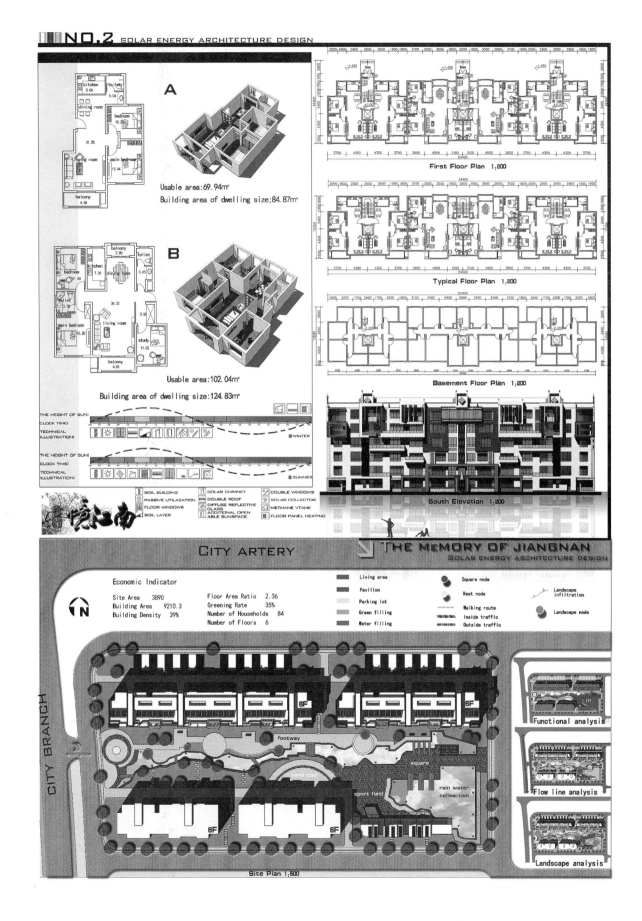

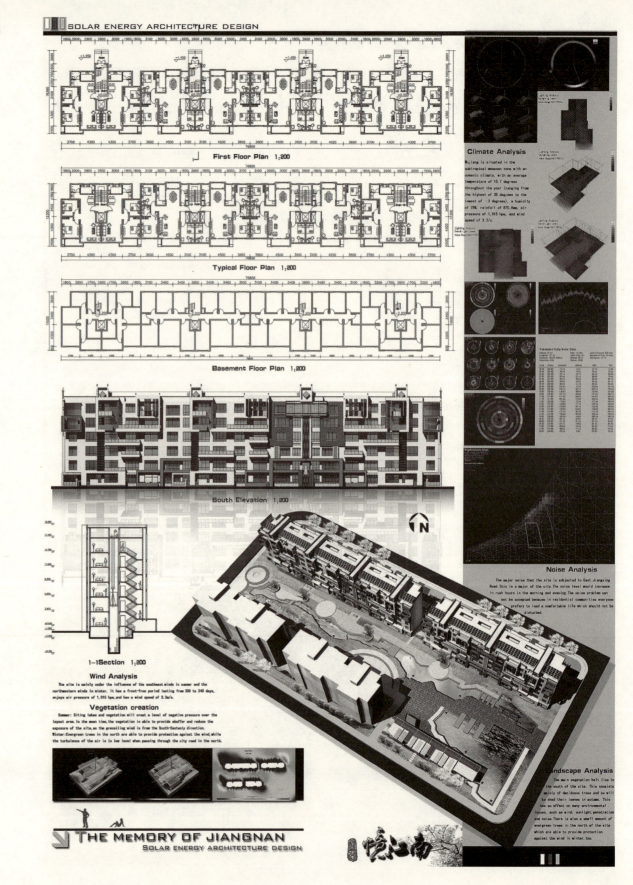

NO.4

THE MEMORY OF JIANGNAN
SOLAR ENERGY ARCHITECTURE DESIGN

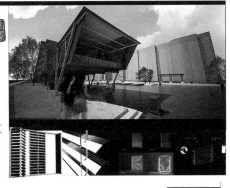

Solar wall

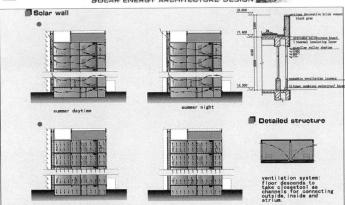

Detailed structure

ventilation system: floor descends to take closestool as channels for connecting outside, inside and atrium.

solarwall: in addition supplying us with fresh air, solarwall can organically deliver heated air to the house.

Shutter shading system

People always used to make the blinds open and close continuously to adjust the angle in order to allow more sunlight shining into the house. It also ensures the shutters on the storage of solar panels to capture more sunlight in the daytime.
The solar battery provide electrode metal sheet on the rollar shutter with electricity power. Thus the room can be lighted up at night.

Double-skin curtain wall construction

summer: According to hot effect, open the bottom vent of the inner wall and the upper air vents of outer wall to discharge hot air in the room.
winter: Open the lower vent of the outer wall and the upper air inlet of the inner wall in, in order to let the fresh air in, which can be used to keep the indoor temperature relatively comfortable, and to block the outdoor noise.

Water recycle system

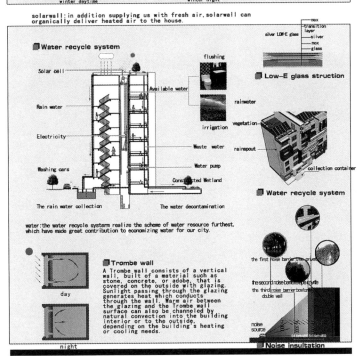

Low-E glass struction

Water recycle system

water: the water recycle system realize the scheme of water resource furthest, which have made great contribution to economizing water for our city.

Solar house

Alternative solar house can be used to collect, store and distribute solar energy in winter, which can be chosen to put off sunshine and exchange fresh air in summer at the same time.

Trombe wall

A Trombe wall consists of a vertical wall, built of a material such as stone, concrete, or adobe, that is covered on the outside with glazing. Sunlight passing through the glazing generates heat which conducts through the wall. Warm air between the glazing and the Trombe wall surface can also be channeled by natural convection into the building interior or to the outside, depending on the building's heating or cooling needs.

Green roof

A green roof is a roof of a building that is partially or completely covered with vegetation and a growing medium, planted over a waterproofing membrane. Also known as "living roofs", green roofs serve several purposes for a building, such as absorbing rainwater, providing insulation, creating a habitat for wildlife, and helping to lower urban air temperatures and combat the heat island effect.

Low-E glass

Noise insulation

PMR Roof Assembly with Membrane Below the Insulation Green Roof Design

Main strategy

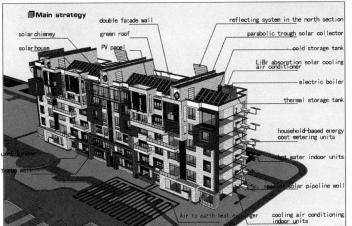

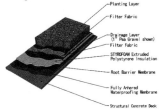

Photovoltaic system analysis

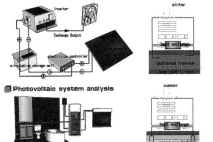

Solar water heating   **Earth source heat pump**

Sunshinewe design solar house for every household; it keeps residents warm in their homes. Meanwhile, install separated-solar water heating, solar air conditioner system photovoltaic system, in order to make full use of solar energy and save energy.

优秀奖
Honorable Mention Prize

项目名称：呼吸的蒙古包
　　　　　Yurt Breathe
作　　者：陈 伟、刘 炳、刘 哲、
　　　　　聂文静、艾尚宏
参赛单位：山东建筑大学建筑城规学院

2627

Hohhot

wind speed　humidity

sun radiant　temperature

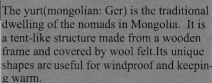

Climate?
typical eastern
subtropical monsoon
climate
summer
hot/rainy
winter
cold/dry

Problems?
keep warm in winter
air in summer
reduce wind speed
solar energy

The yurt(mongolian: Ger) is the traditional dwelling of the nomads in Mongolia. It is a tent-like structure made from a wooden frame and covered by wool felt. Its unique shapes are useful for windproof and keeping warm.

The Monglians are leading a nomadic life together with their indispensable livestock. They are brought up with a character of boldness and unrestrainedness and they have kept a rich and colorful culture.

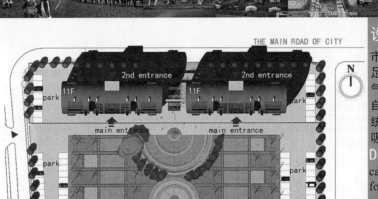

SITE PLAN

设计说明：呼和浩特市低碳宜居住宅设计，立足于当地实际情况，考虑到自然资源和气候条件，注重以太阳能、风能为主的自然资源的结合和利用。并与当地的传统民居——蒙古包结合，创造一个会呼吸的蒙古包。

Design report: The design of the carbon livable residential in Hohhot is founded upon the local resource, environment and climate, focusing on making use of solar energy and wind energy source. We try to combine the design with the traditional yurt, creating a yurt, which can breathe.

YURT ● BREATHE
Low Carbon Livable residential

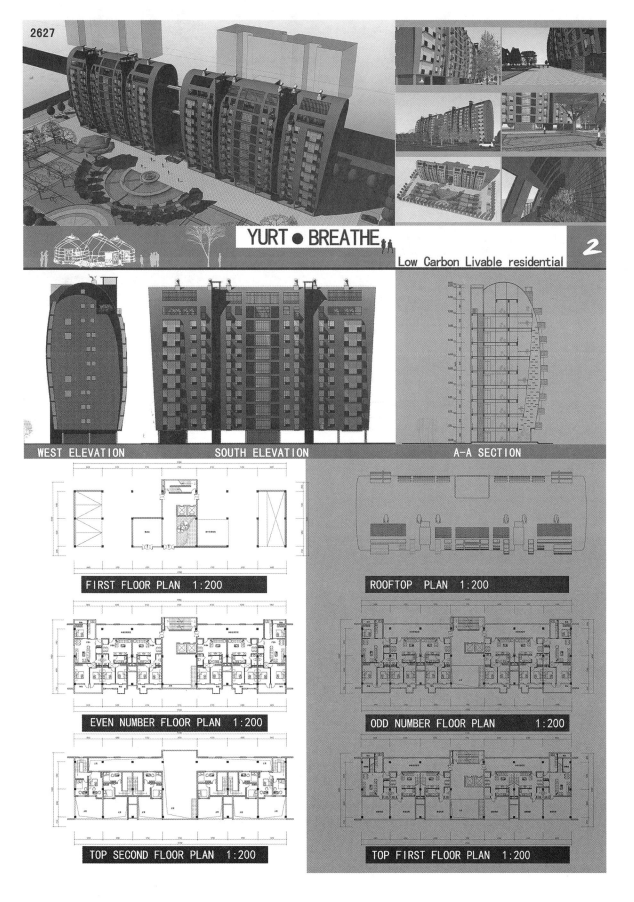

The design not only creates a lifefull form and structure but also brings a special experience of urbanism for people.

Energy strategy—Incorporation

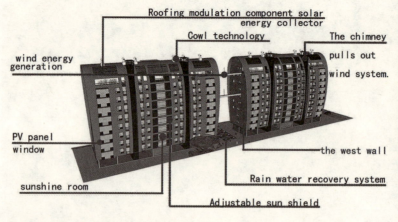

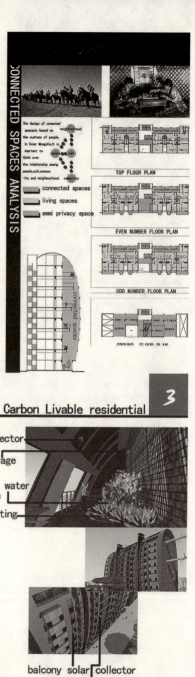

YURT ● BREATHE
Low Carbon Livable residential

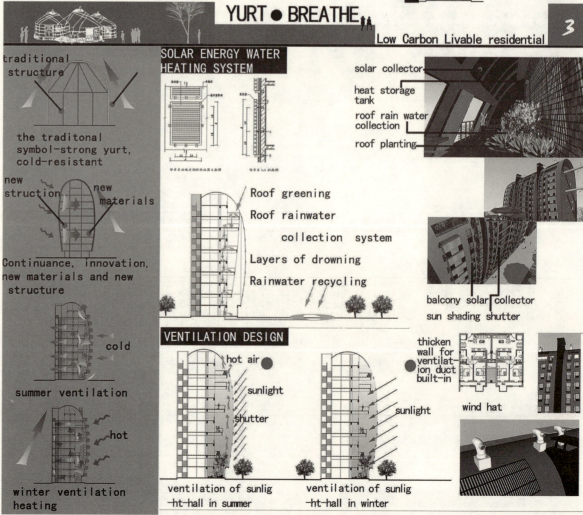

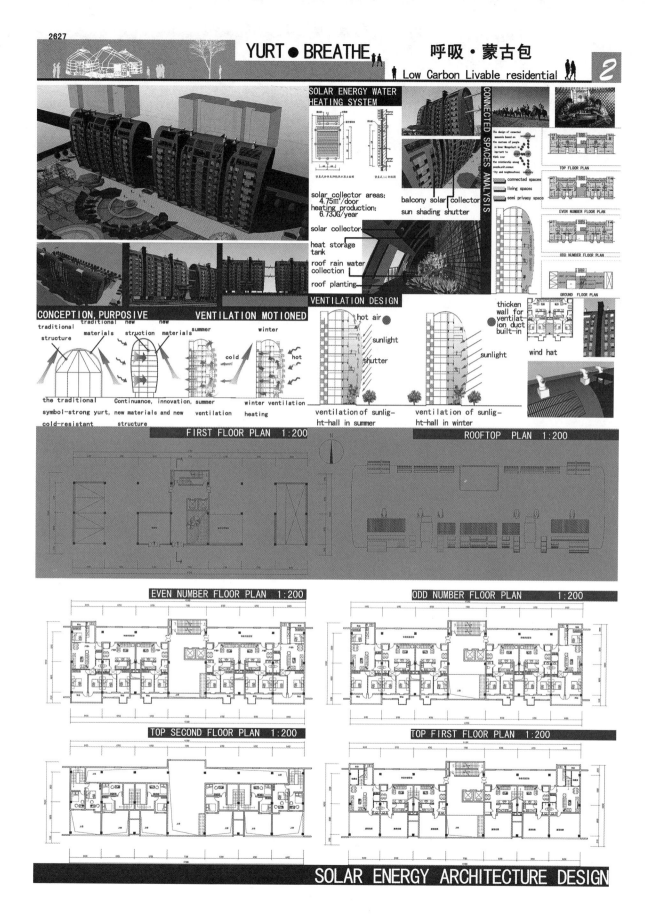

优秀奖
Honorable Mention Prize

项目名称：栖居·间奏
Dwells on This Earth · Rhythm

作　者：赵鹏、赵皞、凌晨、
　　　　李龙飞、朱明瑜
参赛单位：山东建筑大学建筑城规学院

栖居·间奏
Dwells on this earth · Rhythm

同里文化 /Tongli culture
Tongli ancient town with beautiful scenery is surrounded by five lakes. The town is divided into 7 small islands by 15 rivers arranged like the Chinese character "川" connected by 49 ancient bridges, making these islands an integrated area. The architectures built by riverside, famous for "small bridge, flowing stream and households" make Tongli ancient town the best-reserved ancient town by riverside, which has been listed as one of 13 scenicspots of Taihu Lake.

区位分析 /Location Analysis

方案构思 /Scheme Concept
The program is located in Tongli, Jiangsu Province, with humid climate all around the year. In order to reduce the air humidity inside the dwelling and giving occupants a pleasant living environment, our architectural design for the overall planning, environmental layout, house pattern designs started from the residential ventilation. Also we added the concept of low-carbon, solar and other technologies to combine with the local environment.

方案坐落于江苏同里，全年气候湿润，为了减少住宅内部的空气湿度，给居住者一个宜人的居住环境，建筑设计从整体规划、环境布置、户型设计都围绕住宅通风展开。同时加入低碳、太阳能等技术与当地的地域特色相结合。

太阳能综合运用 /Comprehensive Utilization of Solar Energy
In order to take full advantage of the entire residential solar energy resources, the building is designed mainly using passive solar energy, such as the sun space, the solar air system, solar water heaters and roof greenhouses.

整个住宅设计充分利用太阳能资源，建筑以被动式太阳能运用为主，主要运用有阳光间、太阳能新风系统、太阳能热水器、屋顶温室花房等。

Concept 理念

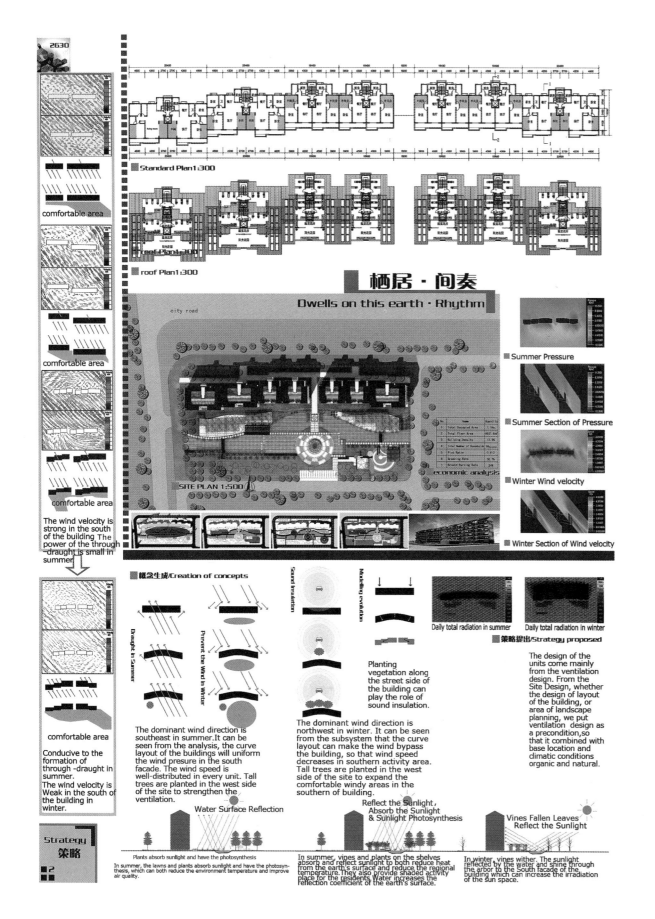

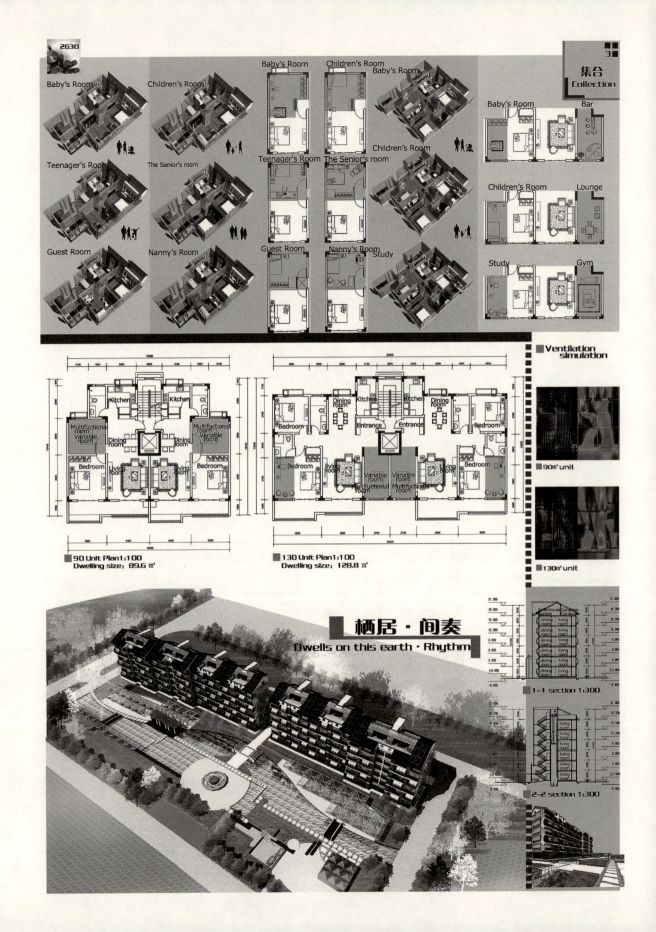

技术 Technology

设计创新/Design Innovation

The design innovation is showed as follow: block insulation system, solar water heater system, overhead vapor barrier system, building self-shading for the north façade, noise reduction for the south façade, roof noise-isolation system, balcony plant irrigation systems and wall ventilation system.

Green metal box + Plant + Maintaining glass
Roof rainwater collection pipe

light rain   moderate rain   heavy rain

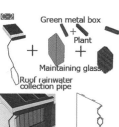

All the rain is absorbed by plants | Most rainwater is absorbed by plants | A few rain is absorbed by plants

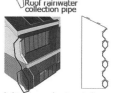

Special-shaped columns makes the layout of the building space more flexible. Since the use of recycled insulation block wall, the building walls become thinner, the thermal insulation and dampproof performance is superior.

Water purification   Water pipe
Water pump   Water intake

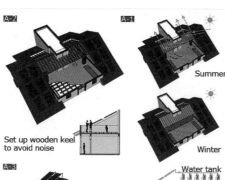
A-2 Summer / Winter
Set up wooden keel to avoid noise

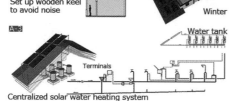
A-3 Water tank
Terminals
Centralized solar water heating system

A 1 Greenhouse
2 Aerial roof
3 Centralized solar water heating system

B 1 Independent solar air system
2 Building self-shading
3 Movable insulation shade
4 The ventilation noise reduction system

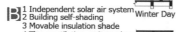

C 1 Greenwall
2 Rainwater irrigation system
3 Shaped frame column
4 Insulation within the external wall

D 1 Ventilation and moisture-proof bottom layer
2 Water source heat pump

C-4 Construction wast   Waste paper box
Recycled concrete hollow block   Paper separator

D-1
a. Surface course
b. Cement plaster damp course
c. Insulating layer
d. Screed coat
e. Cement plaster
f. Building - block
g. Aluminum mesh
h. Air duct & baffle plate

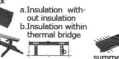
a. Insulation without insulation
b. Insulation within thermal bridge
summer   Winter

LOW CARBON

B-1
a. Awning blind   c. Ecological plant wall
b. Folding window   d. Solar fresh air box

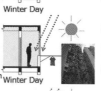
Warm air / Sunlight / Cold air

Winter Day   Summer Day

Winter Day   Summer Day

Summer Night   Winter Night

The traditional way of drying clothes   The method of drying clothes of this building

B-2 The blinds is connected by the connecting rod. They can be turned on at the same time. It is easy use and manage.

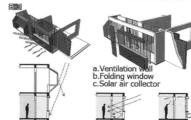
B-3
a. Ventilation wall
b. Folding window
c. Solar air collector

B-4
a. Plant wall
b. Wire meshes
c. Louver
d. Acousitc board

Noise / Quite / Fresh air / Noise / Quite

e. Ventilation window
f. Muffler box
g. Acoustic foam
h. Ventilation window blinds
i. Multihole muffler box

北立面图/North facade 1:300

2011 台达杯国际太阳能建筑设计竞赛获奖作品集

优秀奖
Honorable Mention Prize

项目名称：垂直阳光
Vertical Sunshine

作　者：李俐功、张　良、王　龙、
　　　　王　妍、张　萌、刘彩祥
参赛单位：山东建筑大学建筑城规学院

注册号：2631

1 垂直阳光 VERTICAL SUNSHINE
LOW-CARBON & LIVABLE
吴江市低碳宜居住宅

地理气象 Geography

- **Location**
  Suzhou, jiangsu Near to Shanghai North to old town Tongli
- **Climate**
  Northern semi-tropical monsoon climate.
  Abundant rainfall. Warm all the year.
- **Site**
  South to main road, noisy. Around by Tongli lake.

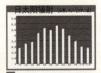

地域文化 Culture

- **Architecture**
  Suzhou Garden. Matou Wall. Elegant buildings.
- **Life**
  Comfort&Easy. Tourism. Low speed.
- **Literature**
  Poem. Kunqu opera.
  Jiangnan literature.

问题思考 Thinking

How to deal with wet weather?
How to regenerate traditional literature?
Can we do more than technology?
What will citizens benefit from buildings?

vol.1

注册号：2631

Technical and economic indexes

total land area 1.1hm²

total construction space 9275.7m²

building density 12.4%

total number of 84

greening rate 36.5%

volume fraction 0.84

ground parking rate 14.3%

2 垂直阳光 | LOW-CARBON LIVABLE
VERTICAL SUNSHINE | 吴江市低碳宜居

1. main bedroom
2. secondary bedroom
3. living room
4. dining room
5. kitchen
6. toilet
7. storage
8. balcony
9. overhaul
10. terrace
11. communicating space
12. working space
13. baby room

Special floor plan 1:200

Technological showcase

- A ventilation apartment door
- B Ventilation stair well
- C openable sun space
- D hollow thermal insulation wall
- E sun shading wooden lath
- F reflecting roof
- G local slate wall
- H indoor gypsum lath insulation wall
- I floor panel heating
- J roof insulation system
- K solar collector
- L anti-western exposure grille
- M ead wind grille
- N roof draining tile

Variable units

As time goes on family's members are changing. Living space should be changed as well to adapt to the needs of different family members.

soho   two-people family   three-people family

vol.2

注册号：2631

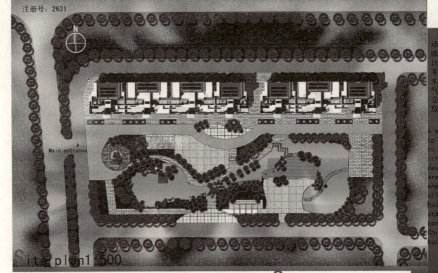

Site plan 1:500

设计说明
Design report

本太阳能建筑设计力图超越技术堆砌，通过场所营造为人们沟通及室外活动创造可能，同时减少对空调的依赖，引领"低碳生活"。总平设计中着重考虑绿化及水面设计，隔噪同时降温，设计考虑经济性及可操作性，技术以被动式太阳能技术为主，从通风、遮阳、降温等几方面回应当地气候条件，尤其利用绿井南作的交流空间兼有社交及生态作用，独具匠心，设计从传统建筑中撷取符号，低调而不流俗地融入古镇风貌。

This design is aimed at being more than a sum of technology and makes it possible to communicate with each other, which leads to less use of air conditioner and "low-carbon life". We think highly of greenland and water design, so that insulating noise and cooling are possible. We rely much on passive solar energy utilization and respond to native climate characteristics with ventilation, sunshade and cooling. Design especially makes stair hall the space for ecology and social contact. Buildings learn much from typical Chinese architecture and live with native buildings in harmony.

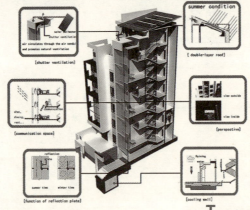

Stair section 1:200

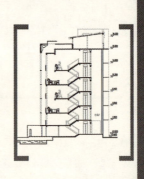

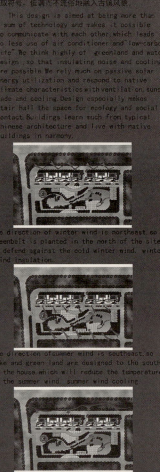

The direction of winter wind is northeast, so greenbelt is planted in the north of the site to defend against the cold winter wind. winter wind insulation.

The direction of summer wind is southeast, so lake and green land are designed to the south of the house, which will reduce the temperature of the summer wind. summer wind cooling.

The greenbelt in the north of the site will insulate the dwelling house from the noise of the north main artery.

Tridimensional virescence

Afforest is decorated in south facade as planting module, it plays the role of shading and decorating the building elevation, removable module can be adjusted as season changes. In west facade we set afforestation plant cavities, in summer plants work as sunshade, in winter greenhouse walls form as leaves wither away, which reduces energy consumption.

3 垂直阳光 VERTICAL SUNSHINE | LOW-CARBON LIVABLE 吴江市低碳宜居住宅

VOL.3

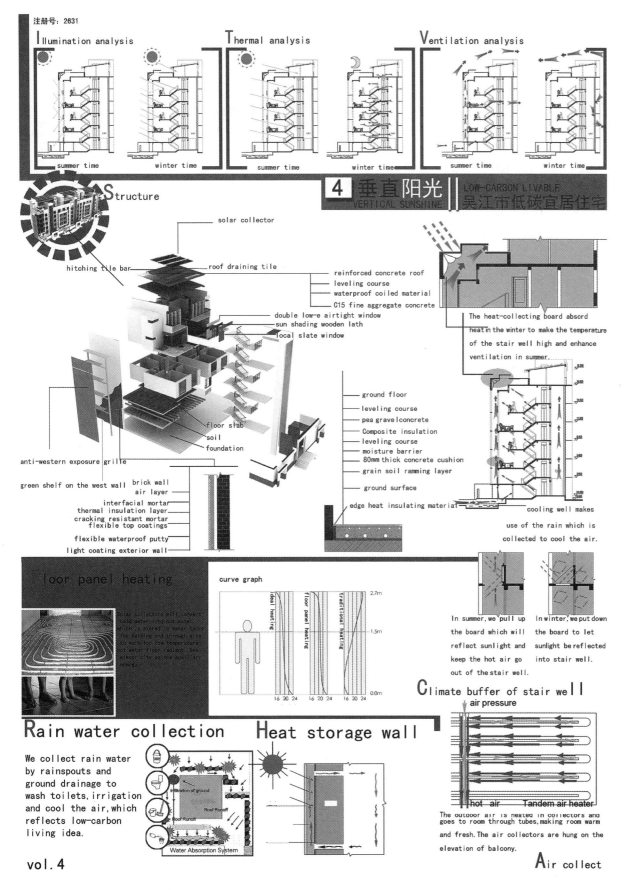

优秀奖
Honorable Mention Prize

项目名称：吴江市低碳宜居住宅设计
　　　　　Change in Sunshine
作　　者：辛　灵、刘　辛
参赛单位：山东建筑大学建筑城规学院

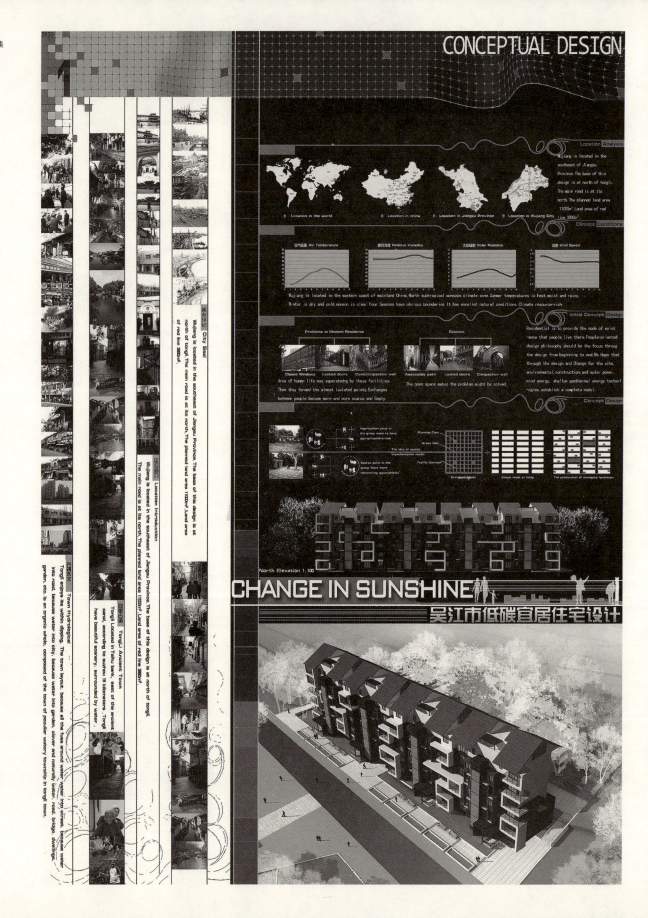

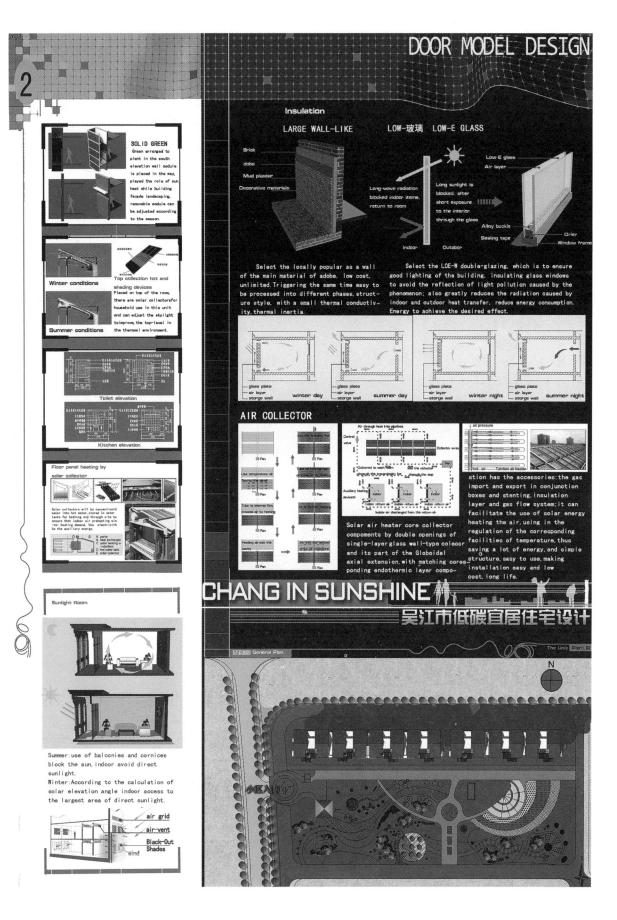

DOOR MODEL DESIGN

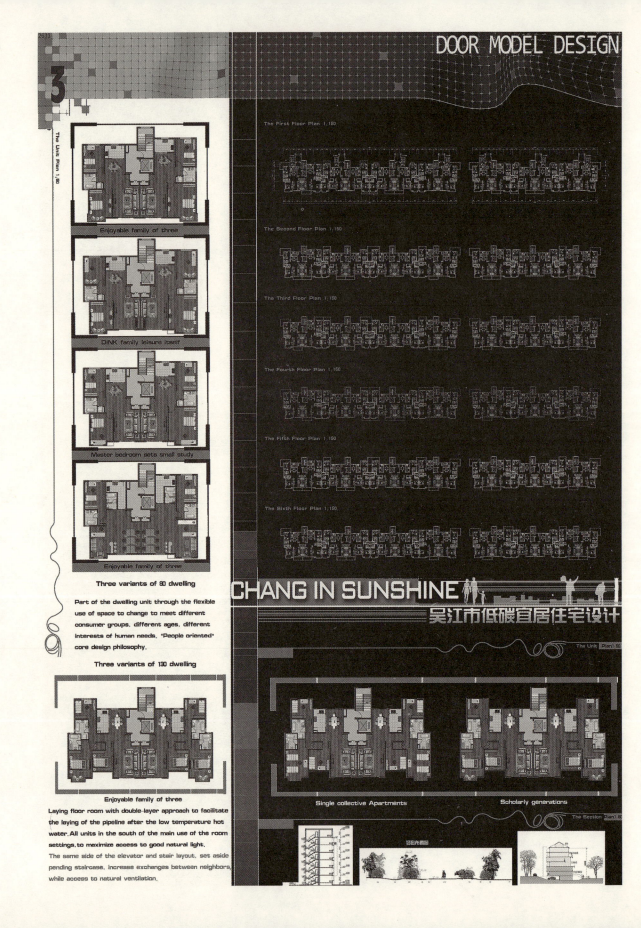

Three variants of 90 dwelling

Part of the dwelling unit through the flexible use of space to change to meet different consumer groups, different ages, different interests of human needs. "People oriented" core design philosophy.

Three variants of 130 dwelling

Laying floor room with double-layer approach to facilitate the laying of the pipeline after the low temperature hot water. All units in the south of the main use of the room settings, to maximize access to good natural light. The same side of the elevator and stair layout, set aside pending staircase, increase exchanges between neighbors, while access to natural ventilation.

CHANG IN SUNSHINE
吴江市低碳宜居住宅设计

二、技术专项奖作品
Prize for Technical Excellence Works

技术专项奖
Prize for Technical Excellence

项目名称：Ice-ray Apartment
作　　者：Chanachok Pratchayawutthirat、
　　　　　Chavanat Ratanamahawong、
　　　　　Tassanaporn Rattanakosate、
　　　　　Farahidina Septiantini、
　　　　　Didit Novianto、Davina Iwana、
　　　　　陈宗炎
参赛单位：University of Kitakyushu（Japan）、
　　　　　浙江大学

专家点评：
方案思路独特，造型效果独特，立面由传统建筑中"窗"的肌理演化而来，其采用的全透光性光伏组件与建筑肌理结合较好地实现了光伏与建筑的一体化结合，具有较强的表现力。

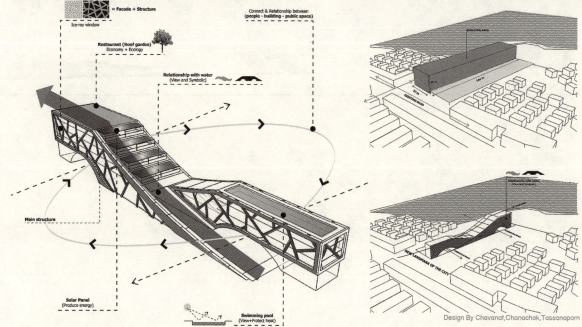

Design By Chavanat,Chanachok,Tassanaporn

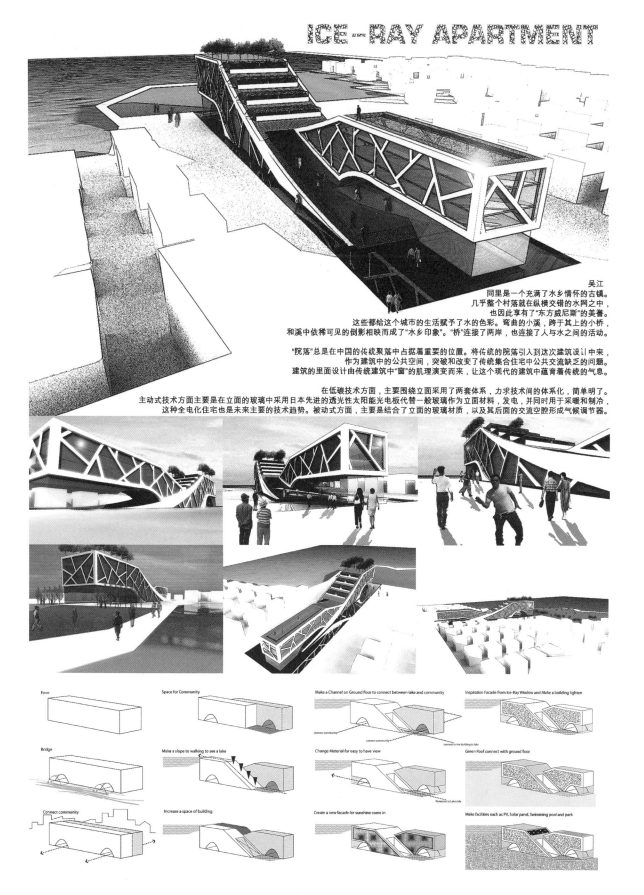

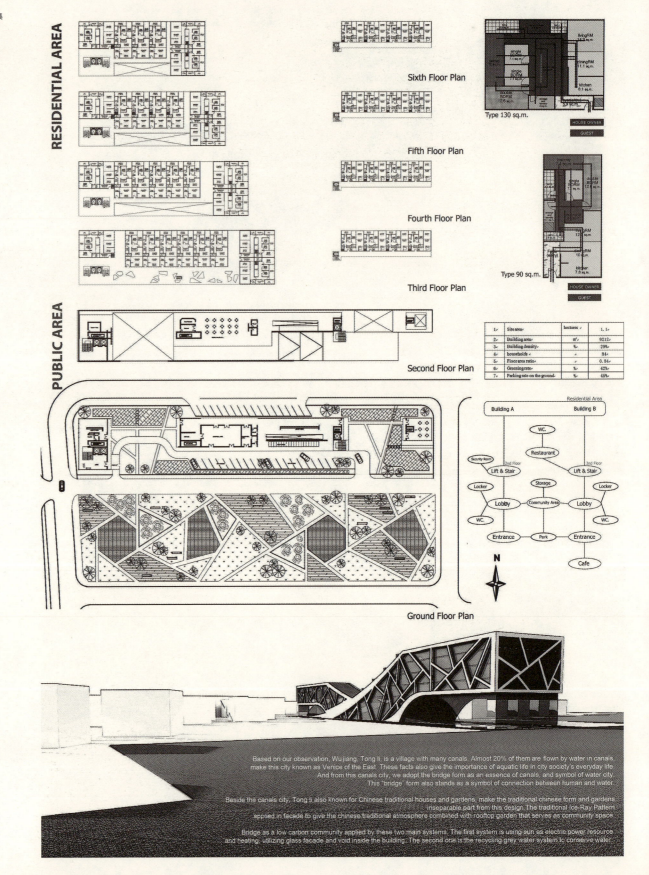

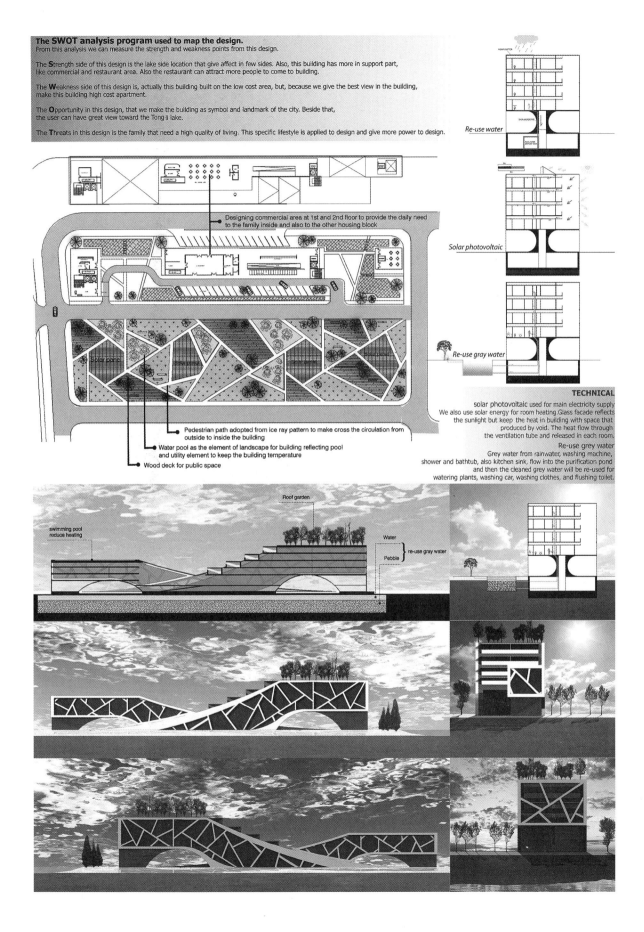

技术专项奖
Prize for Technical Excellence

项目名称：阳光住宅
Solar House

作　　者：任仕新、潘　晖、杨维菊
参赛单位：东南大学建筑学院

专家点评：

方案对绿色建筑材料在建筑上的应用进行了多方面的尝试，与建筑结合较好，并具有一定的可行性。

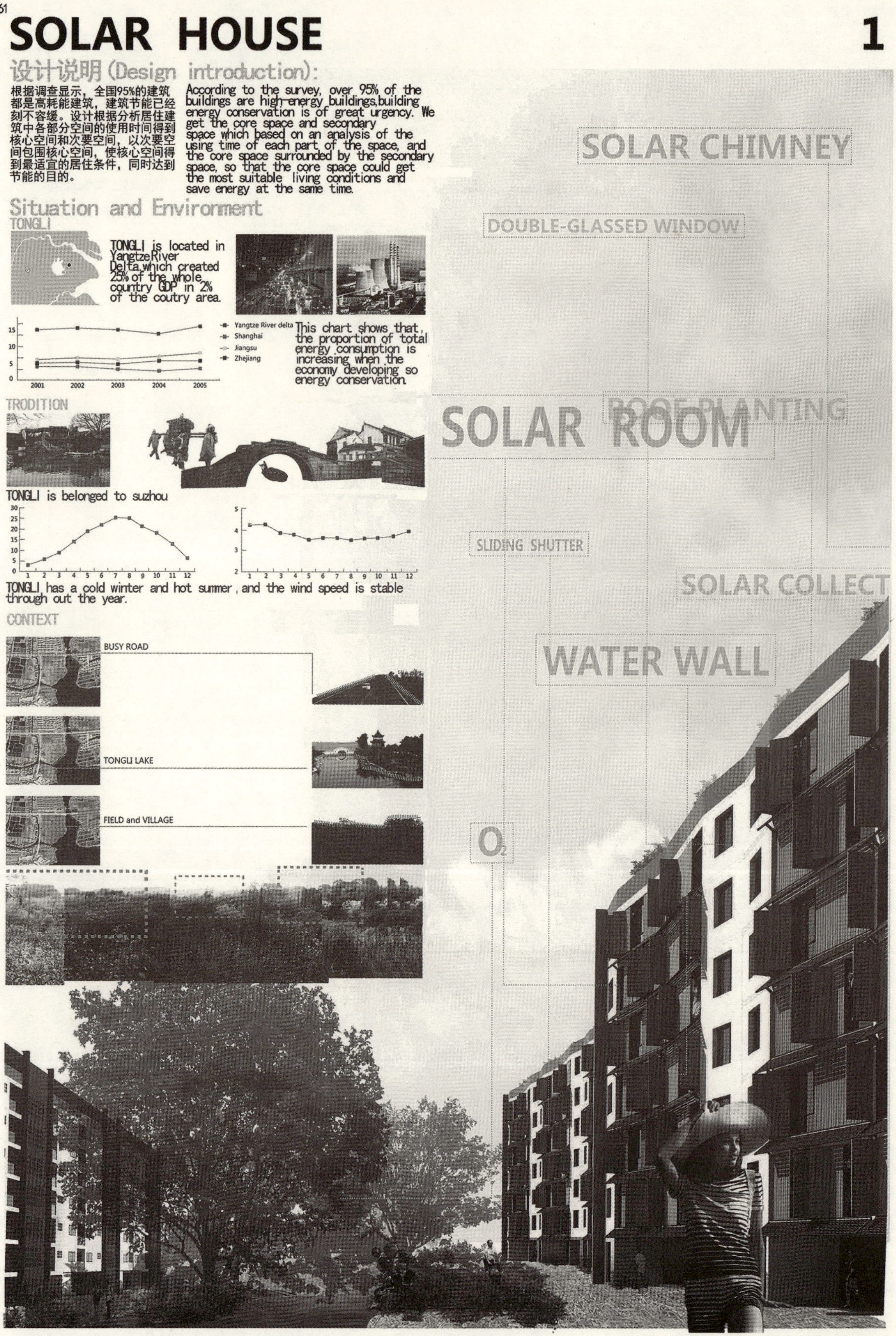

2261
SOLAR HOUSE
1
设计说明(Design introduction):
根据调查显示，全国95%的建筑都是高耗能建筑，建筑节能已经刻不容缓。设计根据分析居住建筑中各部分空间的使用时间得到核心空间和次要空间，以次要空间包围核心空间，使核心空间得到最适宜的居住条件，同时达到节能的目的。
According to the survey, over 95% of the buildings are high-energy buildings,building energy conservation is of great urgency. We get the core space and secondary space which based on an analysis of the using time of each part of the space, and the core space surrounded by the secondary space, so that the core space could get the most suitable living conditions and save energy at the same time.
Situation and Environment
TONGLI
TONGLI is located in YangtzeRiver Delta,which created 25% of the whole country GDP in 2% of the coutry area.
Yangtze River delta
Shanghai
Jiangsu
Zhejiang
This chart shows that, the proportion of total energy consumption is increasing when the economy developing so energy conservation.
TRODITION
TONGLI is belonged to suzhou
TONGLI has a cold winter and hot summer, and the wind speed is stable through out the year.
CONTEXT
BUSY ROAD
TONGLI LAKE
FIELD and VILLAGE
SOLAR CHIMNEY
DOUBLE-GLASSED WINDOW
ROOF PLANTING
SOLAR ROOM
SLIDING SHUTTER
SOLAR COLLECT
WATER WALL
O2

SOLAR HOUSE

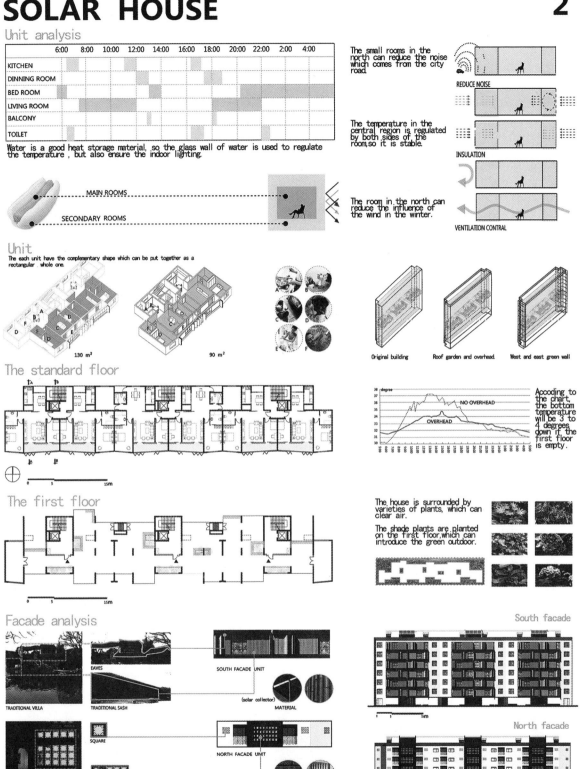

SOLAR HOUSE

The master plan

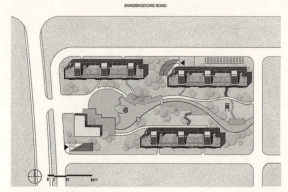

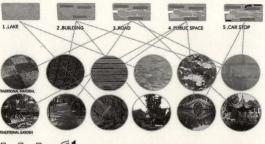

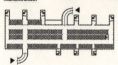

1. Community groups is a reference layout of Suzhou garden style, its material selection is also based on the garden.
2. The parking is located in the community groups underground, which can reach all households directly.

The sections

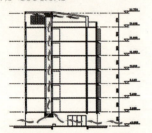

A-A section 1:200

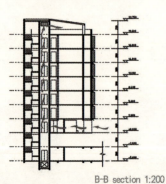

B-B section 1:200

We designed the vertical ventilation shaft and overhead ground to pull air cooling by chimney effect.

Ventilation and thermal environment

As the outdoor temperature in summer and winter can not satisfy the human thermal comfort, we designed a vertical ventilation system which contacts with solar water heaters and the surface cold source to improve the indoor thermal environment.

SOLAR WATER HEATER

WINTER — change — SUMMER

INSTALLATION ANGLE

Installation angle of solar water heaters should be considered carefully, to consider factors such as latitude, water usage in winter and summer and so on.

After comprehensive consideration, we decided to install the solar water heaters to 36 certain angle.

SOLAR WATER HEATER

WINTER SUMMER

OUTLET WOODEN BLINDS

Solar room

Principle

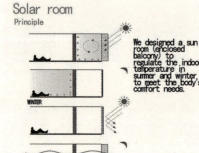

WINTER

SUMMER

We designed a sun room (enclosed balcony) to regulate the indoor temperature in summer and winter, to meet the body's comfort needs.

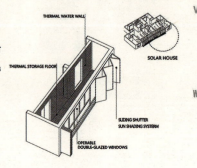

THERMAL WATER WALL

THERMAL STORAGE FLOOR

SOLAR HOUSE

SLIDING SHUTTER SUN SHADING SYSTEM

OPERABLE DOUBLE-GLAZED WINDOWS

Venetian

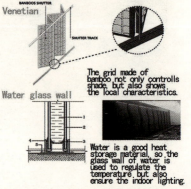

BAMBOOS SHUTTER

SHUTTER TRACK

The grid made of bamboo not only controlls shade, but also shows the local characteristics.

Water glass wall

Water is a good heat storage material, so the glass wall of water is used to regulate the temperature, but also ensure the indoor lighting.

SOLAR HOUSE

Analysis of special details

Plant roof

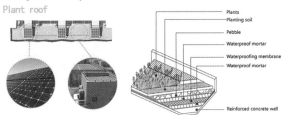

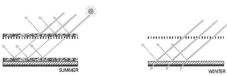

We designed roof gardens, which provide the playing venue while also play a role in summer cooling. We designed racks and sub-surface vegetation in the garden, which rich the garden environment, and provide the better cooling.

Wind folding screens

Winter wind speed is too large in the overhead layer, so it is necessary to set windshields.

INSULATING GLASS

Windows are a major reason for the loss of heat, so we use hollow glass.

GREEN WALL

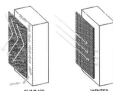

According to the study, the plants on west facade can make the temperatue 4 to 5 degree down.

INSULATING WALL

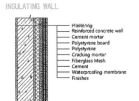

We designed a exterior insulation system which uses the Owens Corning board.

Perspective

技术专项奖
Prize for Technical Excellence

项目名称：光·的容器
　　　　　Solar·Vessel
作　者：陈洸锐、夏伟
参赛单位：清华大学建筑学院

专家点评：
方案巧妙地利用南向阳台的折面处理，使得太阳能集热器、光伏组件能够取得最佳的日照角度，提高了系统效率。

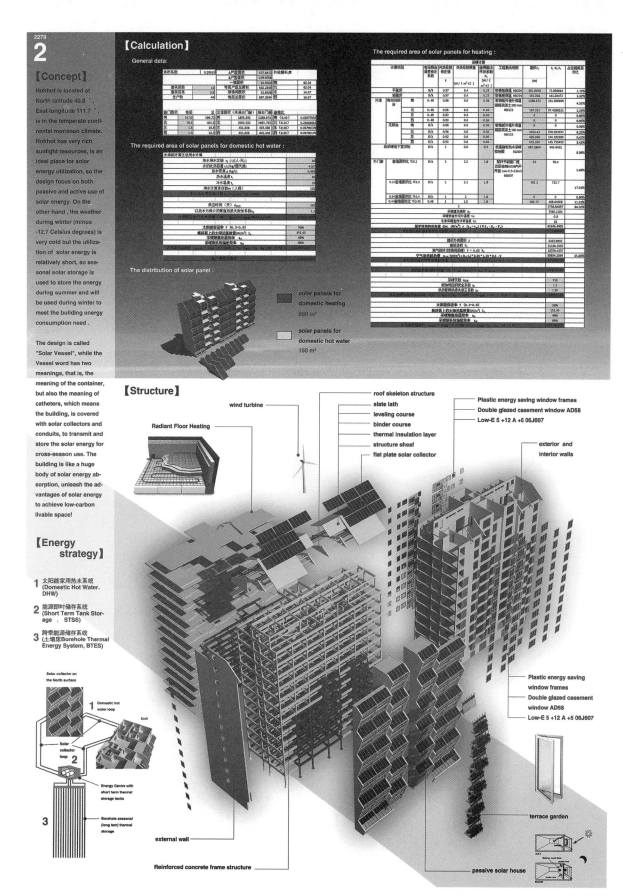

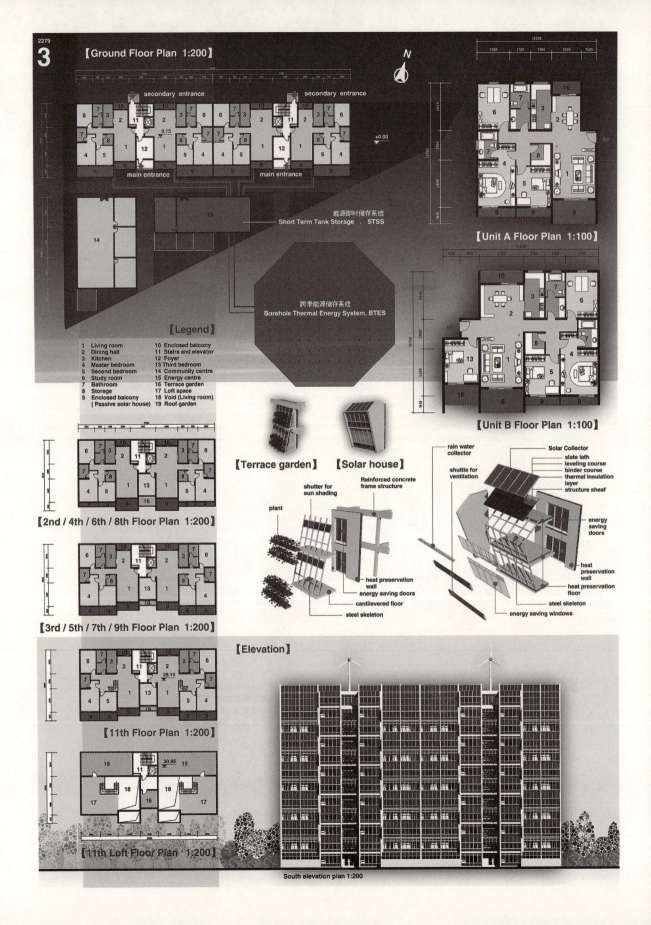

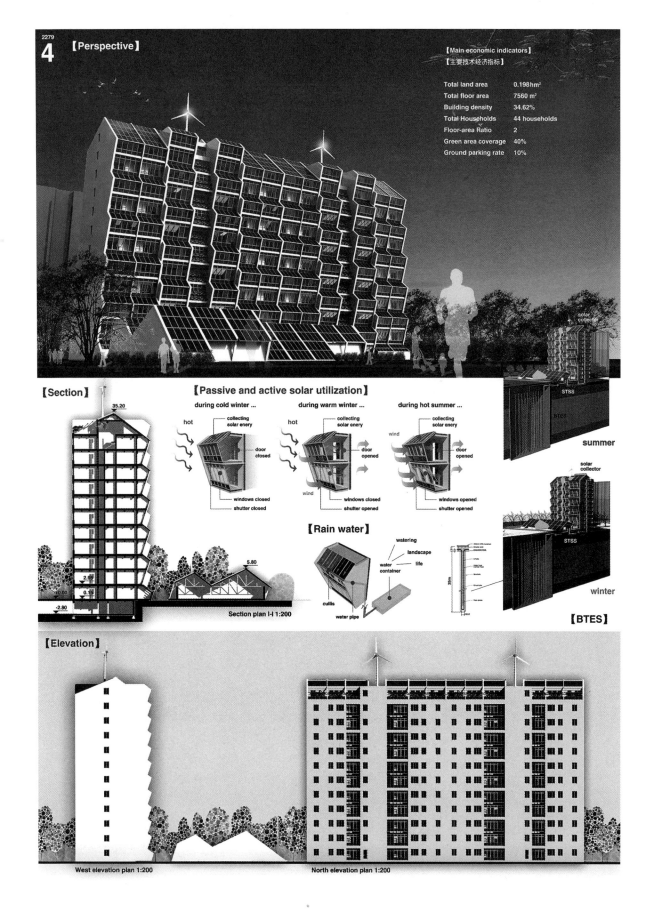

三五间舍 ——吴江市低碳宜居住宅设计

注册号：2557

阳光·低碳

技术专项奖
Prize for Technical Excellence

项目名称：三五间舍
Three or Five Houses
作　者：郑　峥、安　旭、马腾飞
参赛单位：河北工业大学、
　　　　　　沈阳建筑大学、
　　　　　　天津城市建筑学院

专家点评：
方案将传统元素提炼简化并与现代风格相结合，太阳能技术设施与建筑有机结合，造型富有江南水乡特点。

Main View Perspective

Node Perspective I

Node Perspective II

Solar Building DESIGN

设计说明：
该方案选址于苏州地区吴江市，毗邻历史文化名城"同里古镇"和同里湖，周边水系秀美，环境怡人。在构思过程中发掘其江南水乡环境特色，从规划布局到单体设计融入了水乡民居元素，尤其在建筑造型的处理和场地水环境的塑造上，将传统元素提炼简化并与现代风格相结合，展现新江南水乡建筑风貌。同时采用了多种先进的太阳能技术、中水循环处理技术、地热能利用技术、智能通风技术，实现住宅建筑的低碳宜居性。方案立意三五间舍，词句源于宋书营造法式，为表达对中国古代建筑的探究和新环境下传统民居形式发展的思考。

Group Bird's-eye View

2011 Delta-cup International Solar Building Design

01

三五間舍

注册号：2557

——吴江市低碳宜居住宅设计

阳光 低碳

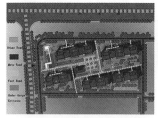
Roads and Traffic System

Landscape System

Ventilation System

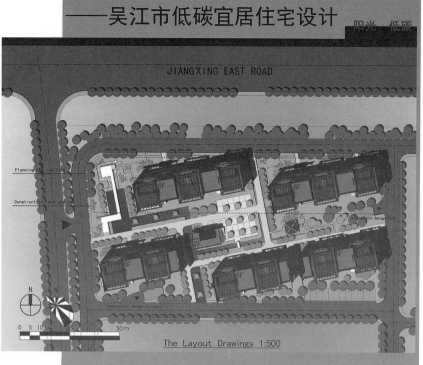
The Layout Drawings 1:500

Solar Building DESIGN

Main technological and economic indicators

Total land area:	1.1h㎡
Total construction area:	9505㎡
Building density:	23.6%
Plot ratio:	0.86
Total family numbers:	144
Surface car parking ratio:	8%

Community groups by the windmill shape of walk from the main road and consists of two ecological water with space frame, organize a car journey of underground car parks near the entrance of setting focus, through pedestrian-and-vehicle dividing system for residents to create a quiet security space.

02

Node Perspective III Node Perspective IV Node Perspective V

South Elevation 1:200

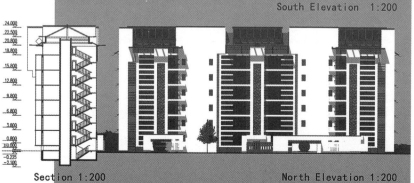
Section 1:200   North Elevation 1:200

2011 Delta-cup International Solar Building Design

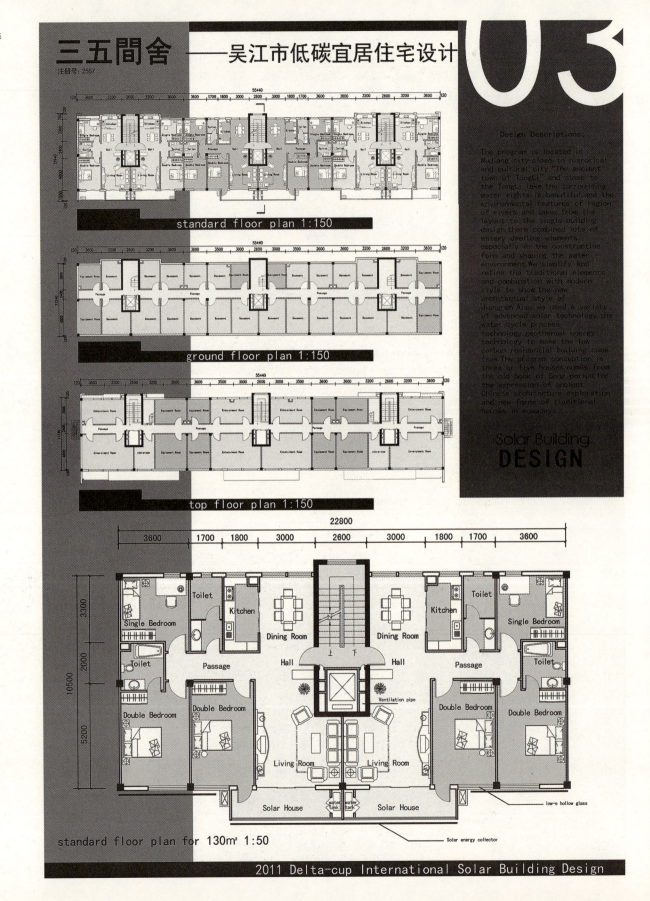

04 三五间舍——吴江市低碳宜居住宅设计

注册号：2557

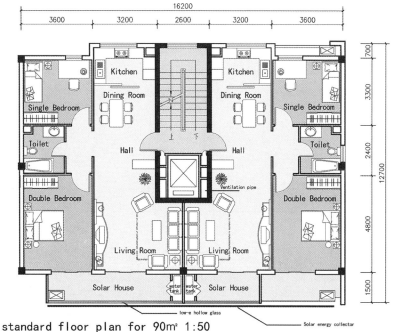

standard floor plan for 90m² 1:50

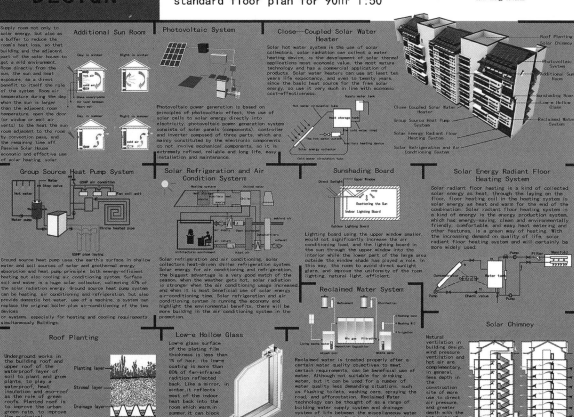

2011 Delta-cup International Solar Building Design

参赛人员名单
Name List of Attending Competitors

注册号	单位名称	作者
1109	东南大学建筑学院	张伟、吴悠、杨维菊
1143	东南大学建筑学院	顾雨拯、季鹏程、杨维菊
1153	中国矿业大学力学与建筑工程学院建筑学系	张方晴、陶如、马全明、张一兵、田海鹏
1196	University of Kitakyushu(Japan)、浙江大学	Chanachok Pratchayawutthirat、Chavanat Ratanamahawong、Tassanaporn Rattanakosate、Farahidina Septiantini、Didit Novianto、Davina Iwana、陈宗炎
1198	重庆市设计院	姜涵、刘学生、张振华、宋东明、黄方艾、游畅
1224	江苏省盐城市工学院	殷义杰、唐燕萍、邓清
1225	江苏省盐城市工学院	孙正、邵泽彪、钱海斌
1226	江苏省盐城市工学院	胥旗、嵇德坤、王贵东
1228	江苏省盐城市工学院	许益亮，李禹希、周荣春
1229	江苏省盐城市工学院	万金翔、邓清
1230	江苏省盐城市工学院	万金翔、陈洋、邓清、黎晓龙
1242	江苏省盐城市工学院	解越、严海明、杨小飞
1256	北京斯维克可持续发展工程设计研究院	刘昌励
1260	大连理工大学建筑与艺术学院	张瑞娜、刘鸣、范悦、袁杰、李莉、陈滨、张宝刚、段苏桐
1370	西安理工大学	崔弘扬
1374	河北工业大学	刘凯
1382	西华大学建筑学院	王行健、秦瀚
1399	盐城工学院	卞铭尧、龚俊程、李阳
1402	东南大学建筑学院	李欣、江雯、杨维菊
1412	盐城工学院	李磊、王晓骏、孙佳乐
1419	盐城工学院	陈利利、孙程
1420	盐城工学院	袁君、崔远奇
1424	天津市建工集团建筑设计有限公司	王彪
1426	盐城工学院	姚政、张健、姜丽雯

续表

注册号	单位名称	作者
1430	东南大学建筑学院	李晓东、岳文昆、朱堃、武鼎鑫、夏晨晨、黄莹
1435	上海大学建筑系	陈曦、倪江涛、王辰君、王润栋、郑康奕、刘裕辰
1441	兰州理工大学	柯熙泰、郑驰浩、张善林、李晶
1444	盐城工学院	卞竹青、黄凯、姜丽丽
1446	青岛理工大学建筑学院	林梁
1470	华中科技大学建筑与城市规划学院	徐燊、陈晗、张林琦、王博
1472	哈尔滨工业大学建筑学院	邵腾、赵丽华、陈琳、王鸿洋
1473	哈尔滨工业大学	陈曦、张莹、侯拓宇、吕环宇
1499	中原工学院	付品强
1508	东南大学建筑学院	王明、杨文杰、王陶
1512	天津大学建筑学院	高辉、董文亮、王梓、石莹、项瑜、盖凯凯、王华峰、王晋、欧阳文
1524	东南大学建筑学院	张良、张瑞文、李蕊
1539	石家庄铁道大学建筑与艺术学院	许炤斐、杨晓、周玉
1550	华中科技大学建筑与城市规划学院	吴璨、陈彦君、朱月、颜艳
1561	山东建筑大学建筑城规学院	魏瑞涵、张晨光、王寅璞、刘滨洋、姜咏茜、杨星晰
1563	盐城工学院	曹辰辰、朱敏敏、秦璐娟
1593	浙江树人大学城建学院	谢永平、许杰、孙飞龙、邵云清、王修水
1599	华中科技大学	奥托卡
1603	沈阳建筑大学建筑与规划学院	任乃鑫、蒋文杰
1604	山东工艺美术学院	王德志
1607	山东建筑大学建筑城规学院	李亮、陶然、康玉东、刘杰民、马淑洁
1610	河北大学建筑工程学院	穆毅、宋晓强、张博、关毅鹏
1616	石家庄铁道大学建筑与艺术学院	高力强、刘恋、沈纪超、王帅、欧阳仕淮、宋宏浩、张袆娇、徐禛龙
1617	浙江树人大学城建学院	吕扬伟、何海峰、宋廉、王修水
1620	浙江树人大学城建学院	徐泽铭、金丽敏、徐剑锋、王修水

续表

注册号	单位名称	作者
1628	西安建筑科技大学	李旭鑫、孙逊、刘钊、高欢、吕栋
1638	华中科技大学建筑与城市规划学院	徐燊、叶天威、郝铭、郑前、张立名
1639	大连理工大学建筑与艺术学院	徐辰、于天怡、黄晓芳、刘鸣、范悦、陈滨、路晓东、柯勋
1655	沈阳建筑大学建筑和规划研究生学院	刘瑞芳
1658	东南大学建筑学院	张宏、赵虎、周静、仇怡嘉、敖雷
1660	浙江树人大学城建学院	张日、黄杉杉、苏平、陈波、叶尚晶、俞立卫、黄裕镯、王修水
1678	华中科技大学建筑与城市规划学院	徐燊、廖维、殷实
1679	攀枝花学院	杨昆、梁文耀
1684	河北大学建筑工程学院	马春件、檀松
1744	临沂大学工学院	隋成祥
1747	东南大学建筑学院	郑恒祥、郑一林
1763	浙江理工大学建筑工程学院	龚臣涛、徐晓霞、俞丽媛、方野
1781	华南理工大学建筑学院	谢君琳、段晓宇、贾佳一
1785	西安建筑科技大学	王芳、孟丹、张扬、周书兵、王雪松
1795	武汉大学城市设计学院	黄欣、张璐璐、陈清怡
1807	北京工业大学建筑与城市规划学院	吴江滨、陈大鹏
1814	北京工业大学建筑与城市规划学院	朱艳婷、黄玉洁
1815	清华大学建筑学院	15120003400
1816	南阳师范学院土木建筑工程学院	李静、何明星
1820	清华大学建筑学院	周显坤
1822	上海大学	丁铭铭、陈玥、朱煜霖、杨泽俞、左惠铭
1828	石家庄铁道大学建筑与艺术学院	高力强、张祎娇、徐禛龙、王帅、姚瑶、姚辰明、刘恋、沈纪超
1838	河北联合大学建筑工程学院	王文瑞、牛璐
1843	清华大学建筑学院	李佳婧
1863	沈阳建筑大学建筑与规划学院	陈瑜

续表

注册号	单位名称	作者
1866	沈阳建筑大学建筑与规划学院	廖江宁
1872	内蒙古工业大学	王旭鸣、王慧琪、杭晨琛、王玉冰、李彩凤、陶淑娟
1878	东南大学建筑学院	许杰、王婧如、崔慧岳、高坤、杨维菊
1882	华中科技大学建筑与城市规划学院	方孙韡、吴耀华、刘晖
1883	长安大学建筑学院	方楠、巩河杉、王正凯、余文广、杨志成、吴晓东
1884	长安大学建筑学院	刘劲、苏婧婷、张冬冬、赵敬源
1885	清华大学建筑学院	吕帅、彭哲、王浩然
1907	山东大学	余新燚、赵阳、薛超、胥宁、任和
1908	同济大学建筑与城市规划学院	包恺
1911	北京工业大学建筑与城市规划学院	陈超、武凤文、全贞花、赵耀华、尚春鸽、邓云康、吴翔、李琢、聂依依、王钰惜、霍寥然、侯隆澍、李志永、李宁军、李清清、金示哲、张跃华
1928	济南大学土建学院	潘波、房璐璐、葛林、史洋
1933	西安建筑科技大学	张扬、孟丹、王芳、周书兵、王雪松、张柁
1938	清华大学建筑学院	刘梦佳
1941	山西容海城市规划设计院有限公司	张记亮
1945	北京工业大学建筑与城市规划学院	左娜、李健、范双双、周涵滔、陈未
1949	天津大学建筑学院	高辉、欧阳文、盖凯凯、王华峰、王晋、王帅、石莹、王梓、董文亮、项瑜、吕亚军
1956	东南大学建筑学院	任怀新、赵忠超
1957	东南大学建筑学院	肖虎、陈金梁、郑彬
1961	东南大学建筑学院	郑帅、胡良、孙亚伟
1988	西南交通大学建筑学院	王瑞、谷亚兰、莫颖媚、王国栋
1999	华南理工大学建筑学院	胡南江、蔡宁、黄祖坚、钱乔峰、袁小雨、张智、龚哲、蒋钧海、张宇峰、蔡健
2001	山东建筑大学建筑城规学院	孙倩倩、任娜娜
2012	石家庄铁道大学建筑与艺术学院	高力强、欧阳仕淮、宋宏浩、王帅、张祎娇、徐祯龙、姚瑶、姚辰明

续表

注册号	单位名称	作者
2013	内蒙古工业大学建筑学院	王鹏、康硕、冀倩茹、刘慧梅、王芳芳、李晓庆、李丽娟、刘铮
2014	南京工业大学建筑学院	江文婷、蔡权、刘磊、孙浯
2021	大连理工大学建筑与艺术学院	吕忠正、冯娇、孙琪、张可为
2022	哈尔滨工业大学建筑学院	曲大刚、王无忌
2030	上海大学	徐琳琳、姜瑞清、宋江
2033	华中科技大学建筑与城市规划学院	程婧婧、刘旭明、古震宇、尹航、黄凯
2034	华中科技大学建筑与城市规划学院	虞愿、梅瀚、熊博文
2039	广州大学建筑与城市规划学院、广东肇庆学院生命科学学院	唐龙、周挺、仲勇、王正、余展腾
2047	广东省华南理工大学	陈靖敏、周超然
2051	青岛理工大学建筑学院	张国萌、王辛、杨清、邹玮、李欣、钱城、赵琳
2053	山东省德州市建筑规划勘察设计研究院	李瑞、王超
2056	内蒙古工业大学	邹景初、张鹏、王欣、王旭鸣、陶淑娟
2067	山东建筑大学建筑城规学院、青岛理工大学建筑学院	曲文昕、张子涛、雷阳、刘巢、曲文晓
2069	大连理工大学建筑与艺术学院	徐跃家、梁辰、胡英
2092	华南理工大学建筑学院	刘伟庆、王鹤、徐映恺、祖延龙、侯雅静
2103	沈阳建筑大学建筑与规划学院	任乃鑫、杨磊、刘泽霖
2106	山东建筑大学建筑城规学院	赵腾飞、赵娜、许晓炜、李卓然、刘宇、房文娟
2108	山东建筑大学建筑城规学院	黄海、刘微微、马宾、罗芬兰、牟晓阳、陈琨
2113	山东建筑大学建筑城规学院	张永娇、任娜娜、蔡嘉星、史克信
2128	山东建筑大学建筑城规学院	曹璇、秦瑶、王甦、韩梦薇、王巧雯、王萌
2134	华南理工大学建筑学院	陈嘉健、李楠、李倩、刘倩妮
2153	东南大学建筑学院	罗佳宁、邵如意、丛勐
2154	东南大学建筑学院、中国矿业大学	姚刚、董凌、王玉、栾奕、姜蕾

续表

注册号	单位名称	作者
2168	沈阳建筑大学建筑与规划学院	任乃鑫、杨金金、翁小平
2186	武汉科技大学	潘昊、张愚峰、赵广颖、斯振彬、严露、叶枫
2197	华南理工大学建筑学院	吴杰、刘继骁、黄仁彬、邓钢石、刘登伦
2199	北京工业大学建筑与城市规划学院	范润恬、吴鹏飞、陈未
2220	浙江大学城市学院	王震林、郑雪梅、李凌培、陈泽、陈超雷
2234	沈阳建筑大学建筑与规划学院	吕金鑫
2247	浙江大学城市学院	陈超、樊菲、陆麒江、冯方舟、王慧、程美伊
2249	浙江大学城市学院	郭麟、吕扬、周康、戚玲燕、吕思葭、应小宇
2251	石家庄铁道大学建筑与艺术学院	高力强、姚瑶、姚辰明、王帅、刘恋、沈纪超、欧阳仕淮、宋宏浩
2261	东南大学建筑学院	任仕新、潘晖、杨维菊
2266	华中科技大学建筑与城市规划学院	蔡莹、曹家寅、韩干波、袁黎
2268	沈阳建筑大学建筑与规划学院	闫璐
2272	解放军后勤工程学院军事建筑工程系、清华大学建筑学院、广西艺术学院设计学院	葛贵武、林波荣、吴锡嘉、李娴、彭渤
2277	厦门大学	程麟、郑豪
2279	清华大学建筑学院	陈洸锐、夏伟
2287	北京交通大学	刘明、王高远
2291	北京交通大学建筑与艺术系	夏海山、姜忆南、杜晓辉、王佳、刘冬贺、孟璠磊、侯磊
2308	北京交通大学	夏海山、杜晓辉、姜忆南、张宁、刘艺、贾秀娟、张文超、李美华、曹文博
2327	同济大学设计创意学院	徐菲叶
2329	天津大学建筑学院	吴巍、邢茜
2331	重庆大学建筑城规学院	池云彦、吴玉培、吴绍鹏、王雪松、周铁军
2333	重庆大学建筑城规学院	彭卓、杜萌、胡松玮、潘崟
2335	重庆大学建筑城规学院	熊健吾、张淞茜、丁华、周铁军、王雪松
2338	重庆大学建筑城规学院、香港中文大学	魏琪琳、陈露、张旭、李浩

续表

注册号	单位名称	作者
2342	重庆大学建筑城规学院	王超、谢崇实、姚静、曹宇博
2350	石家庄铁道大学建筑与艺术学院	王立、李玉珊、曹海云、王雅兰
2371	东南大学建筑学院	吴欢瑜、王海宁、周海龙
2372	中央美术学院建筑学院	郭建东、李靖
2389	哈尔滨工业大学建筑学院	康俊、向钧达
2394	西安建筑科技大学华清学院	尚高峰、张琪
2400	华中科技大学建筑与城市规划学院	潘仲远
2410	浙江大学城市学院	张璞、陈林锋、陈哲、方一川、李锞
2428	华中科技大学建筑与城市规划学院	李龙、韩梦涛、王君益
2449	沈阳建筑大学建筑与规划学院	段文博、马立、仁乃鑫、安艳华
2450	沈阳建筑大学建筑与规划学院	蔡晶
2460	沈阳建筑大学建筑与规划学院	石椿晖
2491	重庆大学建筑城规学院	钱中源、聂天奋、朱懋育
2521	河北工业大学	陈稀
2522	北京工业大学建筑与城市规划学院	王晓朦
2524	东南大学建筑学院	张弛、曹婷
2530	重庆大学建筑与城规学院	郭倩、何恭亮、章舒
2536	天津大学建筑学院	肖静
2555	山东建筑大学建筑城规学院	于亮、孔雪婷、李辰歌、张潇、李倩、裘超然
2556	XLGD 建筑师事务所	Xavier LAGURGUE、李芬
2557	河北工业大学、沈阳建筑大学、天津城市建筑学院	郑峥、安旭、马腾飞
2564	沈阳建筑大学城市规划与建筑学院	姜楠、李博浩
2576	武汉科技大学城市建设学院	吴伟、田延芳、王仙龙、张豪、饶斯萌、魏子恒
2586	山东建筑大学建筑城规学院	李斌、刘清越、吴林娟、李振、宋祥、鹿少博
2614	清华大学建筑学院	李宛

续表

注册号	单位名称	作者
2619	哈尔滨工业大学建筑学院	吴鹄鹏、刘小妹、李新欣、王若凡
2621	山东建筑大学建筑城规学院	赵林、刘文峰、冷鑫、宗伟、窦瑞琪、殷晓峰
2622	山东建筑大学建筑城规学院	高昊
2623	山东建筑大学建筑城规学院	王瑜、孙鹏飞、宋羽、谷炳辰、李超
2624	山东建筑大学建筑城规学院	范敏莉、刘润东、林滋然、宇茜峥、张琦、徐琦
2625	山东建筑大学建筑城规学院	王超、由金朝、庄丽娜、赵紫龙、王琦、梁静然
2626	山东建筑大学建筑城规学院	李明达、孙琮皓、刘婕、綦放、王峰
2627	山东建筑大学建筑城规学院	陈伟、刘炳、刘哲、聂文静、艾尚宏
2628	山东建筑大学建筑城规学院	裘超然
2629	山东建筑大学建筑城规学院	徐莉丽、高倩倩、徐苗、张新亚
2630	山东建筑大学建筑城规学院	赵鹏、赵皞、凌晨、李龙飞、朱明瑜
2631	山东建筑大学建筑城规学院	李俐功、张良、王龙、王妍、张萌、刘彩祥
2632	山东建筑大学建筑城规学院	邵明垒、岳磊、韩月慧、张永林、朱凯第
2633	山东建筑大学建筑城规学院	辛灵、刘辛
2634	山东建筑大学建筑城规学院	崔爽、程深、马辰、张丽、吴蔚迪、韩佳君
2638	山东建筑大学建筑城规学院	张航、张化坤、杨庚、梁越、张宵、刘鸿斌
2640	山东建筑大学建筑城规学院	李雯、徐悦、于辰龙、安胶、魏瑞涵、安琪

2011台达杯国际太阳能建筑设计竞赛办法
Competition Brief for International Solar Building Design Competition 2011

竞赛宗旨：

地球是人类生存的共同家园，住宅是我们生活的基本场所，太阳能是低碳住宅的永续动力，设计太阳能住宅是倡导低碳生活的第一步。

竞赛主题：阳光与低碳生活
竞赛题目：1．吴江市低碳宜居住宅；2．呼和浩特市低碳宜居住宅。
主办单位：国际太阳能学会
　　　　　中国可再生能源学会
承办单位：国家住宅与居住环境工程技术研究中心
　　　　　中国可再生能源学会太阳能建筑专业委员会
　　　　　中国房地产研究会住宅产业发展和技术委员会
冠名单位：台达环境与教育基金会
评审专家：崔愷：国际建筑师协会副理事、中国建筑学会副理事长、中国国家工程设计大师、中国建筑设计研究院副院长、总建筑师。
Anne Grete Hestnes女士：前国际太阳能学会主席、挪威科技大学建筑系教授。
Deo Prasad：国际太阳能学会亚太区主席、澳大利亚新南威尔士大学建筑环境系教授。
M. Norbert Fisch：德国不伦瑞克理工大学教授（TU Braunschweig）、建筑与太阳能技术学院院长。
Peter Luscuere：荷兰代尔伏特大学（TU Delft）建筑系教授。
Mitsuhiro Udagawa：国际太阳能学会日本区主席、日本工学院大学建筑系教授。

GOAL OF COMPETITION

Earth is the common homestead of human beings. Dwelling house is the basic place we live. Solar energy is an eternal and continuous power to low-carbon dwelling house. Therefore, design of solar dwelling house is the first step for inspiriting low-carbon life.

THEMES OF COMPETITION:

Low-carbon Life with Sunshine.

SUBJECTS OF COMPETITION:

1. Low-carbon and livable dwelling house in Wujiang;
2. Low-carbon and livable dwelling house in Huhhot.

ORGANIZER:

International Solar Energy Society (ISES).
Chinese Renewable Energy Society (CRES).

OPERATOR:

China National Engineering Research Center for Human Settlements (CNERCHS).
Special Committee of Solar Buildings, CRES.
Committee of Housing Industry Development and Technology, China Real Estate Research Association (CRERA).

SPONSOR:

林宪德：台湾绿色建筑委员会主席，台湾成功大学建筑系教授。

仲继寿：中国可再生能源学会太阳能建筑专业委员会主任委员，国家住宅工程中心主任。

喜文华：甘肃自然能源研究所所长，联合国工业发展组织国际太阳能技术促进转让中心主任，联合国可再生能源国际专家，国际协调员。

冯雅：中国建筑西南设计研究院副总工程师，中国建筑学会建筑热工与节能专业委员会副主任。

组委会成员：包括主办单位、承办单位及冠名单位。办事机构设在中国可再生能源学会太阳能建筑专业委员会。

评比办法：

1. 由组委会审查参赛资格，并确定入围作品。
2. 由评委会评选出竞赛获奖作品。

评比标准：

1. 参赛作品须符合本竞赛"作品要求"的内容。
2. 鼓励创新，作品应体现原创性。
3. 设计作品应满足国家有关技术规范规定和满足使用功能要求，建筑技术与太阳能利用技术具有适配性。
4. 作品中应充分体现太阳能利用技术对降低建筑使用能耗的作用，并具有可实施性。
5. 作品应在经济可行、技术可靠的前提下，具有一定的超前性。
6. 作品评定采用百分制，分项分值见下表：

评比指标	指标说明	分值
规划布局与建筑设计	指规划布局、建筑构思、使用功能和建筑创新等方面	40
主动太阳能利用技术	通过专门设备收集、转换、传输、利用太阳能的技术，鼓励创新	10
被动太阳能利用技术	通过专门建筑设计与建筑构造利用太阳能的技术，鼓励创新	30
采用的其他技术	其他新能源利用技术和节水、节材、节地等方面的技术，鼓励创新	10
技术的可操作性	技术的可行性、普及性和经济性要求	10

Delta Environmental & Educational Foundation.

JURY MEMBERS:

Mr. Cui Kai, Deputy Board Member of IUA (International Union of Architects); Vice President of Architectural Society of China; National Design Master and Chief Architect of China Architecture Design & Research Group.

Ms. Anne Grete Hestnes, Former President of International Solar Energy Society and Professor of Department of Architecture, Norway Science & Technology University.

Mr. Deo Prasad, Asia-Pacific President of ISES and Professor of Faculty of the Built Environment, University of New South Wales, Sydney, Australia.

Mr. M. Norbert Fisch, Professor of TU Braunschweig and president of the Institute of Architecture and Solar Energy Technology, Germany.

Mr. Peter Luscuere, Professor of Department of Architecture, TU Delft, the Netherlands.

Mr. Mitsuhiro Udagawa, President of ISES-Japan and Professor of Department of Architecture, Kogakuin University.

Mr. Lin Xiande, President of Taiwan Green Building Committee and Professor of Faculty of Architecture of Success University, Taiwan.

Mr. Zhong Jishou, Chief Commissioner of Special Committee of Solar Building, CRES and Director of CNERCHS.

Mr. Xi Wenhua, Director-General of Gansu Natural Energy Research Institute; Director-General of UNIDO International Solar Energy Center for Technology Promotion and Transfer; expert in sustainable energy field from United Nations, international coordinator.

Mr. Feng Ya, deputy chief engineer of Southwest Architecture Design and Research Institute of China; deputy director of special committee of building thermal and energy efficiency, Architectural Society of China.

MEMBERS OF THE ORGANIZING COMMITTEE:

It is composed by competition organizer, operator and sponsor. The administration office is a standing body in Special Committee of Solar Energy Buildings, CRES.

APPRAISAL METHODS:

1. Organizing Committee will check up eligible entries and confirm shortlist entries.
2. Jury will appraise and select out awarded works.

APPRAISAL STANDARD:

1. The entries must meet the demands of the Competition Requirement.
2. The entries should embody originality in order to encourage innovation.
3. The submission works should meet relevant national technological codes, regulations and the demands of usable function. The building technology and solar

设计任务书及专业术语：（见附件）

1. 附件1：吴江市低碳宜居住宅气候条件
2. 附件2：呼和浩特市低碳宜居住宅气候条件
3. 附件3：低碳宜居住宅建筑设计任务书
4. 附件4：专业术语

奖项设置及奖励形式：

综合奖：获奖作品建筑设计与所选用太阳能技术具有较强的适配性。

一等奖作品 2名　颁发奖杯、证书及人民币50000元奖金（税前）
二等奖作品 4名　颁发奖杯、证书及人民币20000元奖金（税前）
三等奖作品 6名　颁发奖杯、证书及人民币5000元奖金（税前）
优秀奖作品 30名　颁发证书

技术专项奖：获奖作品在采用的技术或设计方面具有创新，实用性强。
建筑创意奖：获奖作品在规划及建筑设计方面具有独特创意和先导性。

技术专项奖及建筑创意奖作品　名额不限　颁发证书

作品要求：

1. 作品应进行用地红线范围内的总平面规划及实施方案建设红线范围的住宅建筑布局和单体设计。
2. 建筑设计方面应达到方案设计深度；在技术应用方面应有相关的技术图纸和指标。
3. 作品图面、文字表达清楚，数据准确。
4. 作品基本内容包括：

4.1 简要建筑方案设计说明（限200字以内），包括方案构思、太阳能综合应用技术与设计创新等。

4.2 应按照附件3中所提供的"主要技术经济指标一览表"形式，编制相关技术经济指标。

4.3 总平面图比例1:500（含场地及环境设计）。

4.4 单体设计：住宅楼各层平面图、外立面图、剖面图比例1:100～1:200（应能充分表达建筑与室内外环境的关系），单元平面图1:50，重点部位、局部详图及节点大样比例自定，以及相关的技术图表等。

4.5 建筑效果表现图1～3个。

4.6 参赛者须将作品文件编排在840mm×590mm的展板区域内（统一采用竖

energy technology should have adaptability to each other.

4. The submission works should play the role of reducing building energy consumption by utilization of solar energy technology and have feasibility.

5. The submission works should be advanced under the preconditions of economic practicability and technical liability.

6. A percentile score system is adopted for the appraisal.

APPRAISAL INDICATORS:

APPRAISAL INDICATOR	EXPLANATION	SCORES
Layout and building design	Including layout planning, design ideas, usage function, architectural innovation and others	40
Utilization of active solar energy technology	Technology concerning collecting, transforming, transmitting and utilizing solar energy by special equipments. Innovation is encouraged	10
Utilization of passive solar energy technology	Technology of utilizing solar energy by special building design and construction. Innovation is encouraged	30
Adoption of other technology	Other technology concerning new energy utilization, water saving, materials saving and land saving. Innovation is encouraged	10
Operability of the technology	Feasibility, popularization of relevant technology and economy demands	10

THE TASK OF BUILDING DESIGN AND PROFESSIONAL GLOSSARY (Found in Annex)

Annex 1: Climate conditions of the low-carbon and livable dwelling house in Wujiang

Annex 2: Climate conditions of the low-carbon and livable dwelling house in Huhhot

Annex 3: Task of building design of the low-carbon dwelling house
Annex 4: Professional Glossary

PRIZES:

GENERAL PRIZES:

Building design and selected solar energy technology must be excellent in adaptability to each other.

First Prize: 2 winners

The Trophy Cup, Certificate and Bonus RMB 50,000 (before tax) will be awarded.

向构图），作品张数应为2或4张。中英文统一使用黑体字。字体大小应符合下列要求：标题字高：25mm；一级标题字高：20mm；二级标题字高：15mm；图名字高：10mm；中文设计说明字高：8mm；英文设计说明字高：6mm；尺寸及标注字高：6mm。文件分辨率100dpi，格式为JPEG或PDF文件。

5．参赛者通过竞赛网页上传功能将作品递交竞赛组委会，入围作品由组委会统一编辑板眉、出图、制作展板。

6．作品文字要求：除4.1 "建筑方案设计说明"采用中英文外，其他为英文；尽量使用附件4中提供的专业术语。

参赛要求：

1．欢迎建筑设计院、高等院校、研究机构、太阳能研发和生产企业等单位，组织建筑、结构、设备等专业的人员组成竞赛小组参加本次竞赛。

2．请参赛人员访问www.isbdc.cn 或 www.house-china.net/isbdc.cn，按照规定步骤填写注册表，提交后会得到唯一的作品编号。一个作品对应一个注册号。提交作品时把注册号标注在每幅作品的左上角，字高6mm。注册时间2010年7月1日~2010年12月1日。

3．参赛人员同意组委会公开刊登、出版、展览、应用其作品。

4．被编入获奖作品集的作者，应配合组委会，按照出版要求对作品进行相应调整。

注意事项：

1．参赛作品电子文档须在2011年3月1日前提交组委会，请参赛人员访问www.isbdc.cn 或www.house-china.net/isbdc.cn，并上传文件，不接受其他递交方式。

2．作品中不能出现任何与作者信息有关的标记内容，否则将视其为无效作品。

3．组委会将及时在网上公布入选结果及评比情况，将获奖作品整理出版，并对获奖者予以表彰和奖励。

4．获奖作品集首次出版后 30日内，组委会向获奖作品的创作团队赠样书2册。

5．竞赛消息公布，竞赛问题解答均可登陆竞赛网站查询。

所有权及版权声明：

1．为了更广泛地推广竞赛成果，参赛者同意竞赛承办单位及冠名单位享有自行刊登、编辑、出版、展览参赛作品的权利，并各自享有自行使用、授权他人

Second Prize: 4 winners
The Trophy Cup, Certificate and Bonus RMB 20,000 (before tax) will be awarded.
Third Prize: 6 winners
The Trophy Cup, Certificate and Bonus RMB 5,000 (before tax) will be awarded.
Honorable Mention Prize: 30 winners.
The Certificate will be awarded.

PRIZE FOR TECHNICAL EXCELLENCE WORKS:
Prize works must be innovative with practicability in aspect of technology adopted or design.

PRIZE FOR ARCHITECTURAL ORIGINALITY:
Prize works must be originally creative and advanced in planning and building design.

The quota of Prize for Technical Excellence Work and Architectural Originality is open-ended. The Certificate will be awarded. In all cases the Jury's decision will be final.

REQIREMENTS OF THE WORK:
1. The work should include layout of general plan within the site enclosed by red line and design of single dwelling house within the site of the implementation scheme enclosed by red line.

2. The work should reach the depth of scheme design level in building design and should be with relevant technical drawings in technology utilization.

3. Drawings and text must be expressed clearly and its data must be mentioned exactly.

4. Basic contents of the work include:

4.1 A brief of the design scheme (no more than 200 words) including scheme concept, general application technology of solar energy, design innovation, etc..

4.2 Relevant technical and economic indicators expressed with a "list of technical and economic indicators" shown as a sample in Annex 3.

4.3 General layout (1: 500, including site and environmental design).

4.4 Single building design: a set of design scheme for each dwelling house includes: building plan of all floors, facade and section (1:100~1: 200, which could fully show the building and its relationship, inner and outer), stairway building plan (1:50, including dwelling size, stairway, pipe well, etc.), detail drawings for key parts and joints (scale is unlimited) and necessary figures or charts.

4.5 Rendering perspective drawing (1-3).

4.6 Participants should arrange the submission into two or four exhibition panels, each 840 mm×590 mm in size (arranged vertically). Font type should be in boldface. Font height is required as follows: title with 25 mm; first subtitle with 20 mm; second subtitle: 15 mm; figure title: 10 mm; design description in Chinese: 8 mm, in English: 6 mm; dimensions and labels: 6 mm. File resolution: 100 dpi in JPEG or PDF format.

5. Participants should send (upload) a digital version of submission via FTP to the organizing committee, who will compile, print and make exhibition panels for all

使用参赛作品用于实地建设之权利。承办单位及冠名单位在使用参赛作品时将对其作者予以署名，同时对作品将按出版或建设的要求作技术性处理。参赛作品均不退还。

2．作者应对所提交作品的著作权承担责任，凡由于参赛作品而引发的著作权属纠纷均应由作者本人负责。

声明：

1．参与本次竞赛的活动各方（包括参赛者、评委和组委），即表明已接受上述要求。

2．本次竞赛的参赛者，须接受评委会的评审决定作为最终竞赛结果。

3．组委会对竞赛活动具有最终的解释权。

4．为维护参赛者的合法权益，主办方特提请参赛者对本办法的全部条款、特别是蓝色字体部分予以充分注意。

国际太阳能建筑设计竞赛组委会
网　　　址：www.isbdc.cn
　　　　　　www.house-china.net/isbdc.cn　www.house-china.net/scsb
组委会联系地址：北京市西城区车公庄大街19号
国家住宅与居住环境工程技术研究中心，太阳能建筑专业委员会
邮　　　编：100044
联　系　人：刘芳　王岩
联　系　电　话：86-10-88327102，86-10-88327099
传　　　真：86-10-68302808
E-MAIL：isbdc2011@126.com　info@house-china.net

entries.

6. Text requirement: The brief of the design scheme (see 4.1) should be in Chinese and English, the others are in English. Participants should try to use the words from the Professional Glossary in Appendix 4.

PARTICIPATION REQUIREMENTS:

1. Institutes of architectural design, colleges and universities, research institutions and research and manufacture enterprises of solar energy are welcome to make competition groups with professionals of architecture, structure and equipment to attend the competition.

2. Please visit www.isbdc.cn or www.house-china.net/isbdc.cn. You may fill the registration list according to the instruction and gain an exclusive number of your work after submitting the list. One work only has one registration number. The number should be indicated in the top left corner of each submission work with height in 6mm. Registration time: 1st July, 2010 – 1st December, 2010.

3. Participants must agree that the Organizing Committee may publish, print, exhibit and apply their works in public.

4. The authors whose works are edited into the publication should cooperate with the Organizing Committee to adjust their works according to the requirements of printing.

ADDITIONAL ITEMS:

1. Participant's digital file must be uploaded to the Organizing Committee's FTP site (www.isbdc.cn or www.house-china.net/isbdc.cn) before 1st March, 2011. Other ways will not be accepted.

2. Any mark, sign or name related to participant's identity should not appear in, on or included with submission files, otherwise the submission will be deemed invalid.

3. The Organizing Committee will publicize the process and result of the appraisal online in a timely manner, compile and print publication of awarded works. The winners will be honored and awarded.

4. In 30 days after firstly published two of publication of award works will be freely presented by the Organizing Committee to the competition teams who are awarded.

5. The information concerning the competition as well as explanation about the competition may be checked and inquired in the website of the competition.

ANNOUNCEMENT ABOUT OWNERSHIP AND COPYRIGHT:

1. In order to widely promote the result of the competition, participants agree that Competition Operator and Sponsor hold the power of publishing, printing, exhibiting and applying the works. When the works are used by Operator and Sponsor, the names of authors will be affixed. In the mean while the works will be properly treated in technical according to the requirements of printing or construction. All of submissions are not returned back.

附件1：
吴江市低碳宜居住宅气候条件

1. 气象参数

北纬31.1°、东经120.6°、测量点海拔高度8m。

月	空气温度	相对湿度	水平面日太阳辐射	大气压力	风速	土地温度	月采暖度日数	供冷度日数
	℃	%	kWh/(℃·d)	kPa	m/s	℃	℃·d	℃·d
1月	3.9	76.50	2.69	102.3	4.2	3.6	433	1
2月	5.4	75.30	3.14	102.2	4.2	5.5	353	6
3月	9.1	75.50	3.33	101.8	3.9	9.6	277	30
4月	14.7	76.20	4.25	101.2	3.7	15.4	114	142
5月	19.2	78.70	4.78	100.8	3.3	19.9	20	287
6月	22.9	83.40	4.58	100.4	3.4	23.6	0	393
7月	26.1	86.20	5.05	100.1	3.5	26.9	0	513
8月	25.6	86.50	4.71	100.3	3.4	26.3	0	494
9月	22.1	82.60	3.99	100.9	3.5	22.4	0	368
10月	17	77.90	3.4	101.6	3.5	17.2	49	220
11月	11.6	76.70	2.81	102.1	3.8	11.5	189	76
12月	6	76.50	2.68	102.4	4	5.7	363	9
年平均数	15.3	79.30	3.78	101.3	3.7	15.6	1798	2539

2. 空调室外空气计算参数

参数	夏季	冬季
空气调节计算干球温度（℃）	34.6	−5
空气调节计算湿球温度（℃）	28.6	—
空气调节计算日均温度（℃）	31.4	—
通风计算干球温度（℃）	32	2
空气调节计算相对湿度（%）	82	75
平均风速（m/s）	3	2.9
风向	SE	NW

3. 冬季采暖室外空气计算参数

冬季采暖室外计算温度−3℃，冬季室外最多风向平均风速2.9m/s，最多风向NW。

2. All authors must take responsibility for their copyrights of the works including all of disputes of copyright caused by the works.

ANNOUNCEMENT:

1. It implies that everybody who has attended the competition activities including participants, jury members and members of the Organizing Committee has accepted all requirements mentioned above.

2. All participants must accept the appraisal of the jury as the final result of the competition.

3. The Organizing Committee reserves final right to interpret for the competition activities.

4. In order to safeguard the legitimate rights and interests of the participants, the Sponsors ask participants to fully pay attention to all clauses in this document, especially some clauses with blue colors.

Organizing Committee of International Solar Building Design Competition

Website: www.isbdc.cn   www.house-china.net/isbdc.cn   www.house-china.net/scsb

Address of Organizing Committee:
Special Committee of Solar Buildings, CRES
China National Engineering Research Center for Human Settlements
No.19, Che Gong Zhuang Street,
Xi Cheng District,
Beijing, China
Post Code: 100044
Contact persons: Ms. Liu Fang and Mr. Wang Yan
Tel: 86-10-88327102, 86-10-88327099
Fax: 86-10-68302808
E-MAIL: isbdc2011@126.com   info@house-china.net

ANNEX1:

Climate conditions of the low-carbon and livable dwelling house in Wujiang

1. Basic climate conditions

North latitude 31.1°, east longitude 120.6°, altitude height on measure point: 8m.

Month	Air temperature	Relative humidity	Daily solar irradiation (horizontal)	Barometric	Wind speed	Land temperature	Heating degree days	Cooling degree days
	℃	%	kW·h/(℃·d)	kPa	m/s	℃	℃·d	℃·d
Jan.	3.9	76.50	2.69	102.3	4.2	3.6	433	1
Feb.	5.4	75.30	3.14	102.2	4.2	5.5	353	6

附件2：

呼和浩特市低碳宜居住宅气候条件

1. 气象参数

北纬40.8°、东经111.7°、测量点海拔高度1190m。

月	空气温度	相对湿度	水平面日太阳辐射	大气压力	风速	土地温度	月采暖度日数	供冷度日数
	℃	%	kWh/(℃·d)	kPa	m/s	℃	℃·d	℃·d
1月	−10.9	54.40	1.92	86.7	1.4	−16.9	896	0
2月	−6.2	45.80	2.9	86.6	1.8	−11.5	678	0
3月	0.7	38.80	4.03	86.3	2.4	−0.8	536	0
4月	9.4	32.50	5.48	86.1	2.7	10.3	258	0
5月	16.4	38.50	6.42	86	2.6	18.9	50	198
6月	20.9	46.90	6.21	85.7	2.2	23.8	0	327
7月	22.8	59.10	5.35	85.7	1.8	24	0	397
8月	20.8	64.10	4.8	86	1.5	20.8	0	335
9月	15.1	59.40	4.38	86.4	1.6	14.8	87	153
10月	7.7	55.10	3.46	86.7	1.7	6.6	319	0
11月	−1.6	52.20	2.21	86.8	1.6	−3.5	588	0
12月	−8.9	55.10	1.76	86.8	1.4	−13	834	0
年平均数	7.3	50.20	4.08	86.3	1.9	6.2	4246	1410

2. 空调室外空气计算参数

参数	夏季	冬季
空气调节计算干球温度（℃）	29.9	
空气调节计算湿球温度（℃）	20.8	—
空气调节计算日均温度（℃）	25	—
通风计算干球温度（℃）	26	
空气调节计算相对湿度（%）	64	56
平均风速（m/s）	1.5	1.6
风向	SSW	NW

Continued Table

Month	Air temperature	Relative humidity	Daily solar irradiation (horizontal)	Barometric	Wind speed	Land temperature	Heating degree days	Cooling degree days
	℃	%	kW·h/(℃·d)	kPa	m/s	℃	℃·d	℃·d
March	9.1	75.50	3.33	101.8	3.9	9.6	277	30
April	14.7	76.20	4.25	101.2	3.7	15.4	114	142
May	19.2	78.70	4.78	100.8	3.3	19.9	20	287
June	22.9	83.40	4.58	100.3	3.4	23.6	0	393
July	26.1	86.20	5.05	100.1	3.5	26.9	0	513
August	25.6	86.50	4.71	100.3	3.4	26.3	0	494
Sep.	22.1	82.60	3.99	100.9	3.5	22.4	0	368
Oct.	17	77.90	3.4	101.6	3.5	17.2	49	220
Nov.	11.6	76.70	2.81	102.1	3.8	11.5	189	76
Dec.	6	76.50	2.68	102.4	4	5.7	363	9
average/y	15.3	79.30	3.78	101.3	3.7	15.6	1,798	2,539

2. Calculation parameters for HVAC system design

Parameters	Summer	Winter
Outdoor air conditioning design dry bulb temperature (℃)	34.6	-5
Outdoor air conditioning design web bulb temperature (℃)	28.6	—
Daily mean air temperature (℃)	31.4	—
Dry bulb temperature of ventilation calculation (℃)	32	2
Relative humidity of air conditioning calculation (%)	82	75
Mean air velocity (m/s)	3	2.9
Wind direction	SE	NW

3. Climate parameters for heating system design in winter

Outdoor air conditioning design temperature: -3.0℃.
Dominated wind direction NW, with mean air velocity 2.9m/s.

ANNEX 2:

Climate conditions of the low-carbon and livable dwelling house in Huhhot

1. Basic climate conditions

North latitude 40.8°, east longitude 111.7°, altitude height on measure point: 1,190m.

3．冬季采暖室外空气计算参数

冬季采暖室外计算温度-5.9℃，冬季室外最多风向平均风速4.5m/s，最多风向NW。

附件3：
低碳宜居住宅建筑设计任务书

1．吴江市低碳宜居住宅

1.1自然条件

该住宅设计用地位于江苏省苏州地区吴江市，历史文化名城"同里古镇"北侧、同里湖西岸。用地北临城市主干道江兴东路。项目规划用地红线范围总面积11000m²，实施方案建设用地红线范围面积3890m²，实施方案总建筑面积≥8800m²。

场地现状图可以从下列网址下载：www.isbdc.cn 或www.house-china.net/isbdc.cn。

1.2基础设施

该地区已建有市政自来水、排水、雨水、天然气、供电及通信系统。

1.3设计要求

（1）在项目规划用地红线范围内进行住宅组团规划设计，在实施方案建设用地红线范围内进行建筑设计。

（2）住宅形式为6层集合式住宅（跃层不计入层数），檐口限高22m，每层2户，设电梯。

（3）套型建筑面积标准为：90m²和130m²两种（套型建筑面积不含阳台面积，阳台面积≤10m²）。每个住宅单元只能选用一种套型。每个楼栋至少包含两个不同的住宅单元。楼栋应布置在方案实施建设用地中。套型、单元和楼栋示意见下图。

（4）套型设计以满足主人日常生活使用为基础，应包括如下空间：起居室、餐厅、2～3个卧室、卫生间、厨房等使用空间。另可酌情设置书房、家务室、储藏室等辅助空间。

（5）套型主要房间设计要求如下：

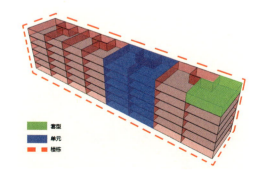

Month	Air temperature	Relative humidity	Daily solar irradiation (horizontal)	Barometric	Wind speed	Land temperature	Heating degree days	Cooling degree days
	°C	%	kW·h/(°C·d)	kPa	m/s	°C	°C·d	°C·d
Jan.	-10.9	54.40	1.92	86.7	1.4	-16.9	896	0
Feb.	-6.2	45.80	2.9	86.6	1.8	-11.5	678	0
March	0.7	38.80	4.03	86.3	2.4	-0.8	536	0
April	9.4	32.50	5.48	86.1	2.7	10.3	258	0
May	16.4	38.50	6.42	86	2.6	18.9	50	198
June	20.9	46.90	6.21	85.7	2.2	23.8	0	327
July	22.8	59.10	5.35	85.7	1.8	24	0	397
August	20.8	64.10	4.8	86	1.5	20.8	0	335
Sep.	15.1	59.40	4.38	86.4	1.6	14.8	87	153
Oct.	7.7	55.10	3.46	86.7	1.7	6.6	319	0
Nov.	-1.6	52.20	2.21	86.8	1.6	-3.5	588	0
Dec.	-8.9	55.10	1.76	86.8	1.4	-13	834	0
average/y	7.3	50.20	4.08	86.3	1.9	6.2	4,246	1,410

2. Climate parameters for HVAC system design

Parameters	Summer	Winter
Outdoor air conditioning design dry bulb temperature (°C)	29.9	
Outdoor air conditioning design web bulb temperature (°C)	20.8	—
Daily mean air temperature (°C)	25	—
Dry bulb temperature of ventilation calculation (°C)	26	
Relative humidity of air conditioning calculation (%)	64	56
Mean air velocity(m/s)	1.5	1.6
Wind direction	SSW	NW

3. Climate parameters for heating system design in winter

Outdoor air conditioning design temperature: -5.9°C.
Dominated wind direction NW, with mean air velocity 4.5m/s.

ANNEX 3:

Task of building design of the low-carbon dwelling house

Low-carbon and livable dwelling house in Wujiang
1.1 Natural conditions
The site for dwelling design is located in Wujiang city, Suzhou region, Jiangsu province, on the north of the historic and cultural town "Tongli Old Town" and at the

a) 单人卧室使用面积不应小于6m², 双人卧室使用面积不应小于10m²。

b) 起居室（厅）使用面积不应小于12m², 应尽量减少直接开向起居厅的门的数量, 起居室（厅）内布置家具的墙面直线长度宜大于3m。

c) 餐厅、过厅如无直接采光, 其使用面积不宜大于10m²。

d) 厨房使用面积不应小于5m², 厨房应设置洗涤池、案台、炉灶及排油烟机、热水器等设施或为其预留位置, 单排布置设备的厨房净宽不应小于1.50m, 双排布置设备的厨房其两排设备之间的净距不应小于0.90m, 套内应设置电冰箱的位置。

e) 卫生间至少应配置便器、洗浴器、洗面器三件卫生设备或为其预留位置。三件卫生设备集中配置的卫生间的使用面积不应小于2.50m², 套内应设置洗衣机的位置。

f) 住宅设计应预留相应的设备管线空间。

（6）新建住宅外观风格需结合同里古镇水乡风貌和当地民居建筑特色。

（7）住宅建设用地红线距北侧城市干道33m, 设计上应采取降低噪声的相关措施。

（8）采用的低碳设计对策和技术应有效率分析并具有可实施性。

2. 呼和浩特市低碳宜居住宅

2.1 自然条件

该住宅设计用地位于内蒙古呼和浩特市新区, 用地北临城市干道。项目规划用地红线范围总面积11000m², 实施方案建设用地红线范围面积3890m², 规划用地红线范围内建筑容积率为2。

场地现状图可以从下列网址下载: www.isbdc.cn 或www.house-china.net/isbdc.cn。

2.2 基础设施

该地区已建有市政自来水、排水、雨水、天然气、供电及电信系统。

2.3 设计要求

（1）在项目规划用地红线范围内进行住宅组团规划设计, 在实施方案建设用地红线范围内进行建筑设计。

（2）住宅形式为9~11层集合式住宅（不含跃层）, 檐口限高36m, 每层2~6户, 设电梯。

（3）套型建筑面积标准为: 60m²和90m²两种（套型建筑面积不含阳台面积, 阳台面积≤10m²）。每个住宅单元应包括以上两种套型。每个楼栋可由一个或多个住宅单元组成。楼栋应布置在方案实施建设用地中。

（4）套型设计以满足主人日常生活使用为基础。应包括如下空间: 起居室、

west bank of Tongli Lake. North of the site is close to Jiangxing East Road, one of main roads of the city. The total area of the project planning site enclosed by red line is 11,000 m² and the area of building construction site enclosed by red line is 3,890 m². The total floor area of the apartment buildings is bigger than 8800 m².

Existing map of the site can be downloaded from www.isbdc.cn or www.house-china.net/isbdc.cn.

1.2 Infrastructures

In this area, municipal water supply, drainage, rainwater, natural gas, power supply and communication system have been built.

1.3 Building design requirements

(1) Layout design of apartment building group are carried out within project planning site enclosed by red line and building design is carried out within the area of building construction site enclosed by red line.

(2) The form of the dwelling is a six-storey apartment building (if a duplex unit is on the top floor, no more storey will be calculated). The height of cornice is restricted no more than 22 m. Two dwelling units (families) are on each floor. Elevator is necessary.

(3) The area standards of two types dwelling size are 90 m² and 130 m² (building floor area of the dwelling unit excludes balcony and the balcony of each unit is less than 10 m²). Each dwelling stairway can only adopt one type of dwelling size. Each apartment building should contain at least two different types of dwelling stairway. The apartment building should be arranged in the area of building construction site. The indication of dwelling size, dwelling stairway and apartment building are shown in below figure:

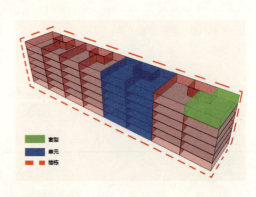

(4) Dwelling size design should meet the requirements of daily life of housemasters which should include following spaces: living room, dinning room, 2-3 bed rooms, toilets, and kitchen, etc.. Besides, it may depend on the condition to arrange study room, housework room, storeroom and other accessorial spaces.

(5) Design requirements of main rooms of the dwelling size:

a) Usable area of a single room is not less than 6 m² while a double room is not less than 10 m².

b) Usable area of a living room (hall) is not less than 12 m². Doors directly facing the living room should be as less as possible. The straight length of the wall of living room (hall) where a set of furniture will be arranged is suitable more than 3 m.

c) Usable area of dinning room and passage hall is not suitable more than 10m² if no direct daylighting.

就餐空间、1~3个卧室、卫生间、厨房、储藏空间等。

（5）套型主要房间设计要求如下：

a）单人卧室使用面积不应小于5m²，双人卧室使用面积不应小于9m²。

b）起居室（厅）使用面积不应小于10m²，应尽量减少直接开向起居厅的门的数量，起居室（厅）内布置家具的墙面直线长度宜大于3m。

c）餐厅、过厅如无直接采光，其使用面积不宜大于10m²。

d）厨房使用面积不应小于4m²，厨房应设置洗涤池、案台、炉灶及排油烟机、热水器等设施或为其预留位置，单排布置设备的厨房净宽不应小于1.50m，双排布置设备的厨房其两排设备之间的净距不应小于0.90m，套内应设置电冰箱的位置。

e）卫生间至少应配置便器、洗浴器、洗面器三件卫生设备或为其预留位置。三件卫生设备集中配置的卫生间的使用面积不应小于2.50m²，套内应设置洗衣机的位置。

f）住宅设计应预留相应的设备管线空间。

（6）住宅建设用地红线距北侧城市干道33m，设计上应采取降低噪声的相关措施。

（7）采用的低碳设计对策和技术应有效率分析并具有可实施性。

3. 主要技术经济指标一览表

序号	名称	单位	备注
1	总用地面积	hm²	
2	总建筑面积		
3	建筑密度	%	
4	总户数	户数	
5	容积率		
6	绿地率	%	≥30
7	地面停车率	%	≥10（停车位与总户数之比）

附件4 / ANNEX 4:

专业术语 Professional Glossary

| 百叶通风 | — shutter ventilation |
| 保温 | — thermal insulation |

d) Usable area of kitchen should be no less than 5 m². In kitchen, a wash basin, an operating table, a cooker unit with hood, water heater, etc. should be set up or their positions should be reserved. The net width of the kitchen is not less than 1.5 m in case all facilities are arranged on one side. For facilities arranged on both sides, the net width of the space between both rows of the facilities is not less than 0.90 m. The position of refrigerator should be arranged in the unit.

e) In toilet, three sanitary devices, toilet bowl, washbasin and bathtub or shower should be set up or their positions should be reserved. Usable area of the toilet where three sanitary devices are set up together should be no less than 2.5 m². The position of washing machine should be arranged in the unit.

f) Spaces for arranging relevant pipes and wires should be reserved in the unit.

(6) The appearance and architectural style of the new buildings are needed to combine with the scene of watery region of Tongli Old Town as well as architectural characteristics of local houses.

(7) The red line of the construction site is 33 m away from the northern main road of the city, the noise prevention should be considered.

(8) The low-carbon design measurement and technology should have efficiency analyses and be feasible.

Low-carbon and livable dwelling house in Huhhot

2.1 Natural condition

The site for dwelling design is located in new zone of Huhhot City, Inner Mongolia, close to a main road of the city from the north. The total area of the project planning site enclosed by red line is 11,000 m², the area of the building construction site enclosed by red line is 3,890 m² and the floor area ratio in the planning site enclosed by red line is 2.

Existing map of the site can be downloaded from www.isbdc.cn or www.house-china.net/isbdc.cn.

2.2 Infrastructures

In this area, municipal water supply, drainage, rainwater, natural gas, power supply and communication system have been built.

2.3 Building design requirements

(1) Layout design of apartment building group are carried out within project planning site enclosed by red line and building design is carried out within the area of building construction site enclosed by red line.

(2) The form of dwelling is a nine-eleven-storey apartment building (if a duplex unit is on the top floor, no more storey will be calculated.). The height of cornice is restricted no more than 36 m. 2-6 dwelling units (families) are on each floor. Elevator is necessary.

(3) The area standards of two types dwelling size are 60 m² and 90 m² (building floor area of the dwelling unit excludes balcony and the balcony of each unit is less than 10 m²). Each dwelling stairway should adopt two types of dwelling size. Each apartment building can contain one dwelling stairway or more. The apartment building

被动太阳能利用	—passive solar energy utilization
敞开系统	—open system
除湿系统	—dehumidification system
储热器	—thermal storage
储水量	—water storage capacity
穿堂风	—through-draught
窗墙面积比	—area ratio of window to wall
次入口	—secondary entrance
导热系数	—thermal conductivity
低能耗	—lower energy consumption
低温热水地板辐射供暖	—low temperature hot water floor radiant heating
地板辐射采暖	—floor panel heating
地面层	—ground layer
额定工作压力	—nominal working pressure
防潮层	—wetproof layer
防冻	—freeze protection
防水层	—waterproof layer
分户热计量	—household-based heat metering
分离式系统	—remote storage system
风速分布	—wind speed distribution
封闭系统	—closed system
辅助热源	—auxiliary thermal source
辅助入口	—accessory entrance
隔热层	—heat insulating layer
隔热窗户	—heat insulation window
跟踪集热器	—tracking collector
光伏发电系统	—photovoltaic system
光伏幕墙	—PV facade
回流系统	—drainback system
回收年限	—payback time
集热器瞬时效率	—instantaneous collector efficiency
集热器阵列	—collector array
集中供暖	—central heating
间接系统	—indirect system
建筑节能率	—building energy saving rate

should be arranged in the area of building construction site.

(4) Dwelling size design should meet the requirements of daily life of housemasters which should include following spaces: living room, dinning room, 1-3 bed rooms, toilet, kitchen, and storeroom, etc..

(5) Design requirements of main rooms of the dwelling size:

a) Usable area of a single room is not less than 5 m^2 while a double room not less than 9 m^2.

b) Usable area of a living room (hall) is not less than 10 m^2. Doors directly facing the living room should be as less as possible. The straight length of the wall of living room (hall) where a set of furniture will be arranged is suitable more than 3 m.

c) Usable area of dinning room and passage hall is not suitable more than 10 m^2 if no direct daylighting.

d) Usable area of kitchen should be no less than 4 m^2. In kitchen, a wash basin, an operating table, a cooker unit with hood, water heater, etc. should be set up or their positions should be reserved. The net width of the kitchen is not less than 1.5 m in case all facilities are arranged on one side. For facilities arranged on both sides the net width of the space between both rows of the facilities is no less than 0.90 m. The position of refrigerator should be arranged in the unit.

e) In toilet, three sanitary devices, toilet bowl, washbasin and bathtub or shower should be set up or their positions should be reserved. Usable area of the toilet where three sanitary devices are set up together should be no less than 2.5 m^2. The position of washing machine should be arranged in the unit.

f) Spaces for arranging relevant pipes and wires should be reserved in the unit.

(6) The red line of the construction site is 33 m away from the northern main road of the city, the noise prevention should be considered.

(7) The low-carbon design measurement and technology should have efficiency analyses and be feasible.

Main technical and economic indicators

Serial number	Item	Unit	Remark
1	Total site area	hm^2	
2	Total floor area of the building		
3	Building density	%	
4	Number of dwellings	dwelling	
5	Floor area ratio		
6	Greening rate	%	≥30
7	Ground parking rate	%	≥10 (ratio between number of parking places to total number of dwellings)

中文	英文	中文	英文
建筑密度	building density	太阳辐射热吸收系数	absorptance for solar radiation
建筑面积	building area	太阳高度角	solar altitude
建筑物耗热量指标	index of building heat loss	太阳能保证率	solar fraction
节能措施	energy saving method	太阳能带辅助热源系统	solar plus supplementary system
节能量	quantity of energy saving	太阳能电池	solar cell
紧凑式太阳热水器	close-coupled solar water heater	太阳能集热器	solar collector
经济分析	economic analysis	太阳能驱动吸附式制冷	solar driven desiccant evaporative cooling
卷帘外遮阳系统	roller shutter sun shading system	太阳能驱动吸收式制冷	solar driven absorption cooling
空气集热器	air collector	太阳能热水器	solar water heating
空气质量检测	air quality test (AQT)	太阳能烟囱	solar chimney
立体绿化	tridimensional virescence	太阳能预热系统	solar preheat system
绿地率	greening rate	太阳墙	solar wall
毛细管辐射	capillary radiation	填充层	fill up layer
木工修理室	repairing room for woodworker	通风模拟	ventilation simulation
耐用指标	permanent index	外窗隔热系统	external windows insulation system
能量储存和回收系统	energy storage & heat recovery system	温差控制器	differential temperature controller
平屋面	plane roof	屋顶植被	roof planting
坡屋面	sloping roof	屋面隔热系统	roof insulation system
强制循环系统	forced circulation system	相变材料	phase change material (PCM)
热泵供暖	heat pump heat supply	相变太阳能系统	phase change solar system
热量计量装置	heat metering device	相变蓄热	phase change thermal storage
热稳定性	thermal stability	蓄热特性	thermal storage characteristic
热效率曲线	thermal efficiency curve	雨水收集	rain water collection
热压	thermal pressure	运动场地	schoolyard
人工湿地效应	artificial marsh effect	遮阳系数	sunshading coefficient
日照标准	insolation standard	直接系统	direct system
容积率	floor area ratio	值班室	duty room
三联供	triple co-generation	智能建筑控制系统	building intelligent control system
设计使用年限	design working life	中庭采光	atrium lighting
使用面积	usable area	主入口	main entrance
室内舒适度	indoor comfort level	贮热水箱	heat storage tank
双层幕墙	double facade building	准备室	preparation room
太阳方位角	solar azimuth	准稳态	quasi-steady state
太阳房	solar house	自然通风	natural ventilation
太阳辐射热	solar radiant heat	自然循环系统	natural circulation system
		自行车棚	bike parking

后记
Postscript

国家住宅与居住环境工程技术研究中心和中国可再生能源学会太阳能建筑专业委员会是国际太阳能建筑设计竞赛的承办单位，倡议将"梦想照进现实"，竞赛作品不应只停留在纸质阶段，而应将获奖作品实地建设起来，使之成为可供观摩、可经受运行检验的建筑实体。这种创新理念不仅将使国际太阳能建筑设计竞赛成果的展示平台登上新高度，更将对太阳能的建筑应用起到积极的推动作用。

正是基于这一理念，2009台达杯国际太阳能建筑设计竞赛在国内率先实现了竞赛获奖作品的实地建设。根据四川省绵阳市涪城区杨家镇小学灾后重建的实际情况，来自山东建筑大学的一等奖作品"蜀光"，经中国建筑设计研究院深化完成施工图设计，由台达电子集团捐赠建设，已于2011年2月投入使用。

"杨家镇台达阳光小学"是一所异地重建的小学，占地面积2.74hm²，建筑面积6570m²，可容纳18个班。学校包括教学楼、办公楼、宿舍楼、食堂、浴室等建筑，设有200m跑道的运动场、篮球场、乒乓球场，并结合场地布置了生态湿地型自然科学园地。

The Competition operators, China National Engineering Research Center for Human Settlements and Special Committee of Solar Buildings, Chinese Renewable Energy Society (CRES) have proposed that "let dream enter reality", which means competition works should not only rest on the paper but being achieved on the ground as a building entity for view and emulation and being able to withstand the test.

Just based on this concept, 2009 Delta Cup – International Solar Building Design Competition took the lead in putting awarded work into construction. According to the actual situation of Yangjia Zhen School in Fucheng District, Mianyang City, Sichuan Province, the awarded work "First light from morning" made by Shandong Jianzhu University which construction design is deeply completed by China Architecture Design & Research group has been accomplished in construction as a contributed project by Delta Electronic Group. Now, it has been taken into use in Feb 2011. "Delta Sunshine Primary School" is a rebuilt one on a new site of 2.74 hm². Its building area is 6,570 m² with 18 classrooms. The school makes up of teaching building, office building, dormitory, canteen, bathroom, etc. It also has playground with a 200m raceway, basketball ground and ping-pong ground. A natural science

竞赛方案　Competition scheme

实施方案　Implementation scheme

台达阳光小学全景　Whole scene of the school

教学楼内庭院　Courtyard

宿舍楼　Dormitory building

食堂　Canteen

架空隔潮层施工现场
Hollow moisture barrier in construction

架空隔潮层外观
Hollow moisture barrier

设置遮阳板的南立面外观
Sunlight shelter on south facade

坡屋面通风隔热外观
Ventilation and heat insulation on the roof

种植隔热屋面
Heat insulation planting roof

架空隔热屋面
Heat insulation aerial roof

太阳能烟囱外观
Solar ventilation duct

太阳能烟囱顶部空腔
Top cavity of solar ventilation duct

设计团队坚持建筑增量成本少、使用舒适度高、运行成本低和可持续运营的低碳校园设计目标，打造融合现代教育理念和先进建筑技术的新型校园空间，通过体验和教育，提升教育理念，促进低碳理念传播。在阳光学校里，体现区域适应性的设计与建造技术，营造安全健康、舒适节能的育人空间，提供高效能、人性化的教学环境。设计充分考虑绵阳地区的气候特点、建筑文化以及校园运行的能耗需求，结合场地条件和不同建筑的使用功能，合理优化规划布局和建筑空间，实现建筑与自然环境有机结合；充分利用太阳能等可再生能源，根据校址地形条件，充分考虑绵阳地区夏季闷热、冬季潮湿的气候特点，实现夏季以通风、隔热、遮阳、隔潮为主，冬季以集热、保温、避风为主的被动设计策略。

在被动设计优先、建筑与自然环境和谐的设计理念指导下，灾后重建的校园不仅建设成为学生学习的场所，更是一个显性物质文化与隐性人文精神和谐共存的精神家园，让教育成为一种自然的存在，成为一种教育思想潜移默化的浸润。"感谢阳光、感恩爱心"，这是阳光小学学生们的心声。在建成的阳

garden plot of ecotype marsh is rightly located on the site.

The design team has made a new schoolyard space with modern educational concept and advanced building technology, thus promoting low carbon concept. Its design aim for low carbon school is more building quantity and less cost, high comfort in use, low cost operation and sustainable management. The design and building technology of sunlight school embodies clime adaptability and creates safe, healthy and energy efficient teaching environment. The designers have given full consideration on the character of climate, building culture and the needs of energy consumption of the school, rationally made overall arrangement and building spaces and realized organic combination between buildings and natural environment. They have fully utilized solar energy and other renewable energy sources and adopted passive way to suit the character of local climate, which is muggy in summer and wet in winter. The main measures are natural ventilation, heat insulation, sunlight shelter and moisture barrier in summer and heat collecting, heat preservation and wind shelter in winter.

The rebuilt school is not only a studying place for students but a coexistent homestead of substantial culture and human spirits. "Thanks to sunlight, thanks to love" is the thinking of the students. Sunlight school brings together architects'

宿舍北内廊热缓冲层
Inner corridor on northside as a buffer of heat

设置采光天窗的教室内景
Classroom with light dormer

室内墙面涂抹除湿涂料
Dehumidified coatings on wall surface inside the room

生态湿地
Ecotype marsh

食堂屋顶安装太阳能集热器
Solar collector on canteen roof

学生的校园生活
Pupils campus life

光小学里，聚焦了全球建筑设计师的目光，也给老师和学生们带来了别有趣味的绿色生活体验。新建筑对人们的影响，最显著的区别在于加强了使用者与环境的友好互动，培养了学生们对节约能源的意识，养成了环保节能的生活方式和可持续的生活态度。让校园建设与乡村教育共同进步，实现"教育发展"与"环境保护"的和谐共生，让可持续发展的绿色新校园，长久地惠及这片土地以及土地上的子孙。

attention of the world and interesting experiment of green life to teachers and students. The new building has strengthened friendly interaction between users and environment, cultivated the sense and life style of energy saving and environmental protection as well as sustainable life attitude. Campus construction and village education will go ahead together, "Education development" and "environment protection" will coexist and sustainable green schools will forever benefit the earth and offspring on the earth.

Competition scheme, implemental scheme and real scenery of the school.